SCIENCE
à la Mode

●‖◉‖◉‖◉‖◉‖◉‖◉‖◉‖ NEW ‖◉‖◉‖◉‖◉‖◉‖◉‖

SCIENCE
à la Mode

Physical Fashions and Fictions

TONY ROTHMAN

PRINCETON UNIVERSITY PRESS
PRINCETON, NEW JERSEY

PUBLICATION HISTORY

"The Garden of Cosmological Delights" first appeared in a slightly shorter version in *Analog*, May 1985.

A somewhat different version of "Metaflation?" appeared under the title "Has Cosmology Become Metaphysical?" in *Astronomy*, February 1987.

"A Memoir of Nuclear Winter" first appeared in a shorter version in *Analog*, November 1987.

"Genius and Biographers: The Fictionalization of Evariste Galois" was published in *The American Mathematical Monthly*, February 1982. New material appears in this edition.

"Geodesics, Domes, and Spacetime" and "The Evolution of Entropy" appear for the first time in this volume.

Library of Congress Cataloging-in-Publication Data
Rothman, Tony.
Science à la mode : physical fashions and fictions / Tony Rothman.
p. cm.
Bibliography: p.
Includes index.
ISBN 0-691-08484-X (lib. bdg. : alk. paper)
1. Science—Philosophy. 2. Scientists—Psychology.
3. Errors. Scientific. I. Title.
Q175.3.R68 1989
501—dc19
88-23848

This book has been composed in Linotron Times Roman with Goudy Bold

Clothbound editions of Princeton University Press books are
printed on acid-free paper, and binding materials are chosen
for strength and durability. Paperbacks, although satisfactory
for personal collections, are not usually suitable for library rebinding

Printed in the United States of America by Princeton University Press,
Princeton, New Jersey

CONTENTS

PREFACE

One of the alternate titles for this book was . . . *And the Bandwagon Rolls On: Critical Essays on Contemporary Science*. Why it was rejected is difficult to say; good titles are largely a matter of instinct. But beginning with ellipses, it was perhaps not declamatory enough to be successful and almost certainly not staid enough for a university press. At any rate, it was scrapped. Yet part of *Bandwagon* still appeals to me: the mournful glance into the distance as the cymbals and trumpets fade is at least an honest admission that nothing said between these covers will have the slightest effect on anything.

Science à la Mode is loosely about bandwagons or fashion in science, as you prefer. I say loosely because the essays making up the collection were written at different times and places and not with the conscious intention of halting any unstoppable objects. In retrospect, however, it is clear that each essay, at least in part, does take issue with a prevailing view. In some cases I may well turn out to be wrong; in other cases I think it's pretty clear that I'll be proved correct. In any case, I suppose it means that I harbor latent iconoclastic tendencies, or maybe not so latent.

I don't apologize for them. I well understand that nonscientists (especially in the United States) tend to regard skeptics as humorless pedants who refuse to take the latest UFO sighting or Dr. Barnard's elixir of youth too seriously. It is a handicap I am willing to live with (though I hope you will find a few rays of humor penetrating these pages). But what my nonscientist friends fail to appreciate is that a scientist's second duty is to be as skeptical as possible—this is why he or she gets paid. A scientist's first duty is to be creative. I assign creativity first place because without it nothing would come into existence, scientific progress included. Unfortunately, creativity is not a skill that runs from nine to five, 365 days a year. New ideas are rare, the mind often needs rest. During those fallow periods a scientist can still approach problems skeptically and this is crucial for the health of the field—not all creatively produced ideas are good, and the bad ones need to be weeded out by merciless criticism. The traits of creativity and skepticism are necessary complements in science. Although excessive skepticism in the earliest stages of theorizing inhibits creativity and often kills potential successes, once a theoretical or experimental result is announced, a scientist's sworn duty is to try to shoot it down as fast as possible. The history of science is littered with the corpses of nonsensical theories and false discoveries. It is sad to think that most of us devote our careers to dead ends. But it is better than the alternative. If we were forced to live in an overpopulated slum dwelling of ridiculous theories and experiments, science would have progressed as far as astrology.

The first commandment of science is therefore "Thou shalt not covet thine own hypotheses," words that will be found at the conclusion of the fifth essay in this book, "A Memoir of Nuclear Winter." Strangely, scientists often violate their own precepts and this fact explains much of the content of *Science à la Mode*. It is easy to jump on the latest bandwagon when your mind is in one of its usual states of hibernation and you don't have any decent ideas of your own. This is particularly true in my own field, cosmology, where definitive experimental tests of theories are slow in coming. So you can calculate to your heart's content without fear of being proved wrong in the near future.

This long-range indeterminacy is exploited in the first two essays to follow, "The Garden of Cosmological Delights" and "Metaflation?" both coauthored with George Ellis. Ellis, I should remark, is known among cosmologists for the standard text, *The Large-Scale Structure of Space-time*, written with Stephen Hawking, for his dozens of seminal papers in the field, and for his healthy iconoclasm, which probably exceeds my own. The two essays were thus a natural collaboration. In "Garden" we examine the evidence for and against the standard Big Bang model of the universe, as well as a few alternatives. In its companion piece, "Metaflation?" we cast a skeptical (but friendly) eye on the popular inflationary model.

Science à la Mode, however, is not only about cosmology. "Geodesics, Domes, and Spacetime" is as much about history as relativity, or perhaps about the relativity of history. "Genius and Biographers: The Fictionalization of Evariste Galois" is entirely about history, and "A Memoir of Nuclear Winter" is about sociology. No, actually they are about the ego of scientists, the primary culprits responsible for the First Commandment's frequent violation.

The scientific ego is a peculiar one. The best scientists are Peter Pans (I fail in this regard), but the naiveté which accompanies perpetual youth can also be pretty unbearable. As one of my mentors wrote (I will not embarrass him by revealing his name) after he read "Genius and Biographers": "I agree with you completely about the cause of the falsification, viz., the narcissistic self-glorification of infant prodigies grown old." And so you find that scientists, whose vision is so acute when it comes to falsifying each other's theories, have large blind spots when it comes to more personal matters.

There is no better example of this, in fact, than the story surrounding "Genius and Biographers." The article sprang from the background research to a play I had written on Galois and the Russian poet Pushkin. I had hardly buried myself in the archives when it became apparent that the standard accounts of Galois' life were more fairy tale than fact. When Leonard Gillman of the University of Texas heard my version at a cocktail party, he asked me to write a piece for the *American Mathematical Monthly* and give a Mathematics Department colloquium. I fulfilled this task, though I was initially hesitant to wage an all-out attack on the transparently sloppy research of my predecessors. I fretted especially about the last section of the article and offered to omit it, writing to *Monthly* editor Ralph Boas with the unsteady plaint, "Don't you think I'm being too harsh on authority?'' To which he replied, ''What's wrong with being harsh on authority?'' So the closing words stand.

But my misgivings were not entirely unwarranted. On the day I was to announce my findings, the weekly departmental calendar read approximately:

4:00 Colloquium—Tony Rothman
The Life of Evariste Galois. A case of historical falsification.

5:00 Riot
After the talk, Rothman will be torn limb from limb for having the effrontery to challenge some of the most cherished stories in the history of mathematics.

I survived. The audience's judgment was favorable. But the jesting tone of the calendar turned earnest in a *New York Times* editorial which appeared after the *Scientific American* version of the article (the editorial is reproduced in this volume as Appendix A to ''Genius and Biographers'').

I became a debunker of fairy tales, a reputation which will probably not be diminished by this book. Ironically, my bookshelves are lined with dozens of fairy-tale collections from around the world. But a good fairy tale tells us something about psychology and about life; a bad one fails in this. As children we have all learned the story of the frog who, upon condition of receiving a kiss for his services, retrieves the Princess's golden ball from a deep well, where it had fallen from her hands. She honorably obliges and they live happily ever after.

No. Read the original Grimm version. The frog demands not a kiss but the right to eat and sleep with his Princess. She consents but runs off to the castle as soon as the ugly frog fulfills his half of the bargain. When the frog appears at the castle to collect his due, only the intervention of the King—who admonishes his daughter to keep her promises—allows the frog to be seated at dinner. Later, when the frog attempts to climb into her bed, the horrified Princess hurls the poor animal against the wall. As he falls to the ground the miraculous transformation takes place and they live (happily) ever after. I do not doubt which version was passed from the lips of the people and which was passed by Queen Victoria's censors.

The moral is that fairy tales are for adults. Scientists, being Peter Pans, prefer the censored versions. More than one reader has written to me wagering that E. T. Bell's famous chapter on Galois in *Men of Mathematics* will outlive my retelling. Fine. At least now there is a choice.

The ego of scientists and the blind spots resulting from it can have implications beyond squabbles over the life of a mathematician dead for 150 years. More and more scientists are now engaging in the popularization of their own fields. I wholeheartedly support this trend for the simple reason that scientists know their turf better than journalists. At the same time, I am disturbed to see a growing gap between the standards upheld by scientists when they face other scientists and the standards they uphold when they face the public. In public it evidently becomes more difficult for scientists to wrestle their egos to the ground, and this leads them to say things they would never try to get away with among colleagues.

Colleagues frequently tell me that the public doesn't care about details and that I should not worry so much about accuracy. I disagree. It is less excusable to be sloppy before the public than among colleagues. Fellow scientists will torpedo leaky arguments immediately. The public will only be misled. And sometimes it can make a difference.

My colleagues argue that I am overly idealistic: the public cannot follow highly technical arguments. True, to which I respond merely that scientists who know what they are talking about should be able to make explanations in nontechnical language. The essays in this book are therefore as nontechnical as I could make them. But they are not totally lacking in technical detail. "You should make things as simple as possible but not too simple" is one of the uncountable aphorisms attributed to Einstein. Whoever did say it has my support. The reader will be required to follow a few involved arguments. In one or two places, for instance Sections 3 and 4 of "The Evolution of Entropy" (which is about metaphor and transmogrification), you may have to skim the boring bits. But not to worry, the main ideas should shine through undimmed.

I am going to leave you now for the essays, but I do not want to leave you with the impression that I am anti-scientist. I love scientists. Sometimes I am one. I merely propose to treat them by the same rules they treat astrologers, historians, and literary critics. You will not find the meditations in *Science à la Mode* directed against scientists any more than you will find them directed against certain consumer advocates. Nevertheless, I hope after reading the book you will hesitate before hopping on the next bandwagon that rolls by. You will in all likelihood not be able to stop it. But at least you can gaze mournfully into the distance as the cymbals and trumpets fade.

Princeton, New Jersey
June 1988

SCIENCE
à la Mode

1 · THE GARDEN OF COSMOLOGICAL DELIGHTS

(WITH G. F. R. ELLIS)

1. FAIRY TALES

Recently, at a New York cocktail party, a young physicist was asked how he made his living and he replied that he was by specialty a cosmologist. While it might be debated whether cosmology constitutes a "living," his host remained undeterred and immediately inquired if it would be possible to make an appointment for a manicure and a haircut. The physicist explained that cosmology is the study of the large-scale structure of the universe and that he—alas—knew very little about nail polish, split ends, and all those other things a cosmetologist presumably deals with. Both the physicist and his host had a good laugh, after which the host meekly retired with a faint "oh," apparently convinced that cosmology was incomprehensible.

A complementary but equally dismissive view was once expressed in a lecture by Nobel laureate Hannes Alfvén who remarked that present cosmological theories have "the character of ancient Indian myths, with turtles standing on elephants standing on. . . . Very beautiful fairy tales."

The cocktail party host and Alfvén expressed two views that characterize quite accurately the nonspecialist's view of cosmology and the theory on which it is based—general relativity. Either it is the most grandiose enterprise imaginable, combining supreme *chutzpah* and unintelligible mathematics, or it is not physics at all but rather esoteric mythology.

It strikes us that there is some truth in both the Cocktail Party

3

View and the Indian Mythology View, but that the discipline of cosmology really falls somewhere in between. Because cosmological theories make many predictions that are not yet testable by experiment (and may never be), they are by their very nature highly conjectural and a fertile ground for speculation. Yet, in some areas, our present theories make remarkably good predictions and are so esthetically pleasing that it is difficult to believe there is not some truth in them.

In this article we are going to speculate. Too often, in the free press, all attention is devoted to the so-called "standard cosmological model" with the tacit assumption that the standard model is correct and we know everything there is to know about the universe. Here, we are going to ask the question, "What if the standard model is wrong? Are there any alternatives?" Indeed there are. Even if there weren't, the case for the standard model would be logically much stronger if we could show that all the alternatives were incorrect. After all, you can't logically conclude the standard model is *the* model if it is the *only* model you've invented. In fact, this is one of the most important reasons for examining other possibilities. But before we tackle these difficult "eccentric" models, let us start the reader on the beginner's path with a review of the old standby, standard Big Bang cosmology.

2. THE STANDARD MODEL: "DULL AS DISHWATER"

We should first explain what a model is. Einstein's equations *do not* specify the universe; rather, they may be considered a general framework within which you can construct many different model universes. These model universes may have absolutely nothing to do with the real one—and usually they don't—but ideally they should represent the large-scale distribution of matter in our universe and the curvature of spacetime caused by that matter. Such curvature is, for instance, manifested in the famous bending of light around the sun and other celestial objects like black holes. (See "Demythologizing the Black Hole" by R. Matzner, T. Piran, and T. Rothman in *Frontiers of Modern Physics*, Dover,

1985). In addition, the model should also describe the history or evolution of the matter in the universe and hence the history of the universe itself. Now, which model is "correct" can be determined only by self-consistency and comparison with the real universe, and this is where experimentalists come in. In our field, experimentalists are usually called astronomers. Of course, you are quite at liberty to throw out Einstein's equations and write your own— some people do—and this procedure brings about the proliferation of even more models.

For about the last twenty years, one cosmological model has carried the title "standard." It also goes by the name of the Friedmann cosmology, or the Robertson-Walker cosmology, and often the Friedman-Robertson-Walker cosmology and occasionally the Friedmann-Lemaître-Robertson-Walker cosmology, depending on which nationalities are disputing priority. (Friedmann was Russian, Lemaître French, Robertson American, and Walker English.) In any case, the FLRW cosmology is the model popularly known as the Big Bang. There are, in fact, any number of Big Bangs, so we will stick with the acronym FLRW when speaking of the *standard* Big Bang.

Before tearing apart the standard model, it is only fair that we explain why most cosmologists are its ardent supporters. First of all, the FLRW Big Bang is the simplest of all Big Bangs and physicists are highly attracted to the Principle of Simplicity. We will explain exactly what we mean by "simplicity" a little later; but because the concepts involved are somewhat abstract, let us start with the more famous and concrete successes of the standard model.

The FLRW cosmology has made two startling predictions. The first of these is that the light isotopes, most importantly helium and deuterium (heavy hydrogen), were formed roughly three minutes after the Big Bang when the universe was extremely hot. You must keep in mind that in the Big Bang picture the universe cools as it expands, somewhat like the expanding freon that cools your refrigerator. When the universe was three minutes old it was *cool* enough so that neutrons and protons could stick together to form deuterium (at higher temperatures the neutrons and protons merely bounced off each other) but *hot* enough so that helium-forming reactions could take place. This occurred at a temperature of about one billion degrees, much hotter than the center of the sun. When the temperature dropped far below one billion degrees this "primordial nucleosynthesis" stopped and, according to the standard model, we should be left with roughly 25% helium by mass and 2×10^{-5} parts deuterium.

It may seem like a miracle that astronomers in fact *do* measure about 25% helium in the real universe, but it is a miracle squared that they also measure something like 2×10^{-5} parts deuterium. This must be counted as a great success of the standard model.

The second prediction of FLRW is that there should exist relict radiation left over from the cosmic fireball, just as gamma rays are left over from a nuclear explosion. For technical reasons the radiation actually seen comes from about 100,000 years after the Big Bang, when the universe became transparent; but, in any case, the radiation also cooled as the universe expanded and should be observable today not as gamma rays or even visible light but as lower-frequency microwaves. Indeed, in 1965 the famous "cosmic microwave background radiation" was discovered by Arno Penzias and Robert W. Wilson at Bell Labs and explained by Robert Dicke's group at Princeton.

Because these two predictions are so decisive, they are often used to compare one cosmological model to another and we will refer to them frequently. Actually, it is so difficult for a model to predict both the light isotope abundances and the cosmic microwave background that most alternative models have been of the Big Bang type. This fact will become more evident as we go along.

Now, we mentioned that the FLRW was the simplest Big Bang model. In order to do useful work, the physicist must translate words like "simple" into mathematical concepts. We will now explain what simple means to a cosmologist. These concepts are, unfortunately, more abstract than helium and microwaves, and the reader is advised at this point to mix a vodka tonic. Lime, please.

The FLRW assumes that at some finite time in the past, the

universe started to expand from a *singular* state of infinite temperature and density. Furthermore, the density of material (say, neutrons, protons, electrons, photons, etc.) is assumed to have been *uniform* throughout the universe and the expansion of the universe is taken to be *homogeneous* and *isotropic*. Let us illuminate some of these terms. A singularity is a point of spacetime where some quantity becomes infinite. In the FLRW universe—alas—virtually everything becomes infinite at the instant of the Big Bang itself, which is thought to have occurred between 10 and 20 billion years ago. If you think a singularity must be a breakdown of sorts, you are absolutely correct. We will have more to say about this later.

The term "isotropic" refers to a system that looks the same in all directions or, in technical language, is "rotationally symmetric." You might imagine yourself standing at the edge of the Grand Canyon and turning around. The abyss before you does not look like the desert behind you, so the area surrounding the Grand Canyon is certainly *not* isotropic. On the other hand, if you stood like a lizard in the middle of the desert, it might very well look the same in all directions, so we would say the desert *is* isotropic.

By contrast, the term "homogeneous" refers to a system that looks the same at any point or, technically speaking, is "invariant under translations." For instance, if the Grand Canyon were idealized as being very straight and of uniform width, we could walk along it and at any point it would look exactly as it had a moment before. *We could not tell we had moved.* Yet, we could still turn around and see the desert, which appears very different from the canyon. So here we have a situation which is *an*isotropic but homogeneous. Since these terms are very important in cosmology, it is best to remember them: *homogeneity* means no change in landscape when one walks; *isotropy* means no change when one spins. See Figure 1.1 to understand that isotropy everywhere implies homogeneity but *not* vice versa. (Philosophical exercise: is life homogeneous?)

Thus, as foretold, the FLRW cosmology is about as simple as one can get. We may visualize the universe to be filled with radiation such as photons, quarks, and neutrinos, as well as more ordinary matter such as protons and neutrons. This material is

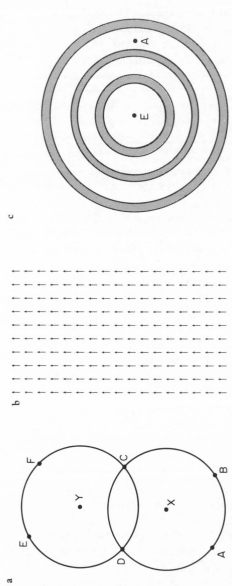

Figure 1.1. (a) Isotrophy implies homogeneity. Suppose the universe is isotropic around point X. Since isotropy means "no change as one spins," then, by definition, points A, B, C, D must all have the same properties. Similarly, if the universe is also isotropic around point Y, then points C, D, E, F must have the same properties. By transitivity A, B, C, D, E, F are all equivalent. Because the universe is the same, having moved from A to B to C to D to E to F, the universe is homogeneous at these points. Extending this argument all over space shows that isotropy everywhere implies homogeneity. (b) Homogeneity does not imply isotropy. Suppose you lived on an infinite field covered with identical arrows all pointing north. If you walked anywhere, it would be impossible to tell you had moved since the landscape has not changed. Hence this universe is homogeneous. But if you look south you see arrowtails, and if you turn north you see arrowheads. There is a preferred direction and so this universe is anisotropic. (These arrows could represent temperature gradients, galactic or particle velocities, or any other property of spacetime.) (c) An inhomogeneous universe. Here the galaxies are not distributed uniformly but concentrated in rings. This universe is clearly inhomogeneous. If you stood at an average point A, it would also appear anisotropic. However, if the observer just happened to be located at point E, the universe would appear isotropic, but only from this point.

absolutely uniform everywhere and in all directions, that is, homogeneous and isotropic. Furthermore, the requirements of homogeneity and isotropy ensure that the universe is expanding at equal rates in all directions and that annoying things like bumps and shock waves do not exist.

Observationally, we cannot actually verify that the universe is homogeneous simply because we cannot travel very far from earth. Even if we could travel 1,000 light years we would still be seeing everything from the same region in our galaxy. Isotropy implies that we cannot point in any particular direction and say "we have seen the center of the universe over there," which is the same as saying, "the universe is very different that way." This isotropy seems to exist approximately in the real universe if we ignore irregularities such as mere galaxies and only consider size scales of galactic clusters and above.

Any cosmological model must predict that the currently observed universe is approximately isotropic. However, we do not call this a "success" of the standard model since it was assumed to be isotropic from the very beginning.

We are going to start multiplying now. The FLRW model itself comes in several styles. The Einstein equations predict the universe is either expanding or contracting and observations of the redshifts of distant galaxies indicate that the universe is presently expanding. (Light becomes redder when emitted from an object moving away from us and bluer when emitted by objects moving toward us. Hence, galactic redshifts indicate the universe is expanding.) The Einstein equations, however, do not specify the amount of radiation or matter present in the model, and these must be determined by direct astronomical observation or other theoretical considerations. If the matter or radiation content of an FLRW universe is below the so-called "critical density," the model will keep expanding forever. In other words the universe is "open." Most evidence indicates that the real universe belongs to this "no frills attached" variety. There are, however, available extras. Massive neutrinos may exist, as well as photinos, gravitinos, Higgsinos, and a host of other new exotic particles which we fortunately cannot discuss here. (Physics has gotten out of hand.) If these particles contribute a sufficient mass density to the universe, the expansion of the universe will eventually halt and the universe will recollapse. Such a universe is often termed "closed."

It is time to ask a stupid question: why is the universe expanding at all? A satisfactory philosophical answer probably can't be given but a physical one can: the universe was born with a certain amount of kinetic energy (energy of motion) and potential energy (gravitational energy). Like a ball being thrown into the air, the universe initially has most of its energy in kinetic energy, but gradually more and more is transferred to potential energy until the ball stops. The ball then has no kinetic energy and falls back to the ground. This is like a closed universe. If, however, the ball has sufficient amounts of kinetic energy, it will reach escape velocity and never fall back to the earth. This is like an open universe.

For the moment, these are all the details we need of the standard model before tearing it apart.

Perhaps the first point that should be made about that standard model is that sixty years ago it would not have been considered standard at all. For philosophical reasons Einstein originally felt that an ideal universe should be neither expanding nor contracting but *static*, and his first cosmological model of 1917 was exactly that. Now, in order to produce a static model of the universe from his equations, Einstein was forced to add the famous "cosmological constant." *This constant may be thought of as adding a term to the potential energy of the universe equivalent to a repulsive force or pressure.* Einstein chose a value for the constant that added just enough so that the kinetic energy of the universe was zero. The ball always "hovered" at the top of its flight. (A more accurate analogy would be a pencil balanced on its point.) Such a situation may seem impossible, and indeed Eddington showed in 1930 that a static universe was unstable and tended to contract or expand. In any case, by that time evidence for the expansion of the universe had been discovered, and in 1931 Einstein dropped the cosmological constant as the "biggest blunder" of his life.

Recently, models with cosmological constants have come back in vogue (see Section 5 of this essay) and are now classed as non-standard models. So the moral of the story is that, like Tchaikovsky and high heels, cosmological models come in and out of fashion. This is the theme of our essay: one should be careful what one calls standard, for tomorrow there may be a replacement. Therefore, the reader would be wise not to forget cosmological constants.

There are more serious objections to the standard model than changing fashion. We mentioned that the FLRW cosmology begins with a singularity. This is a much more serious breakdown than a flat tire or a cracked engine block. It is, in fact, a physical impossibility—a region where the laws of physics break down altogether and even spacetime itself comes to an end. To avoid the singularity is probably the main reason cosmologists search for other models.

There are other conceptual problems with the FLRW Big Bang. Recall that we said it was exactly homogeneous and isotropic. Physicists who follow the Principle of Simplicity are attracted to this model because it is indeed the simplest conceivable cosmology. On the other hand, physicists who rely on the Principle of Greatest Probability (also known as the Principle of Minimum Serendipity) are disturbed. Just how likely is it that the universe was created in an exactly uniform fashion with strict homogeneity and isotropy? Such a birth seems at best implausible but this is exactly what FLRW claims. Doubts such as these led to the creation of anisotropic and inhomogeneous models which we will discuss in Section 3.

There is a further problem with FLRW. If the universe is open, then it is truly infinite in extent. There may have been a Big Bang fifteen billion years ago, but one must regard the Big Bang not as occurring at one point but all over space, not as one pebble dropped in a pond but a pebble plopped at each point. The Big Bang took place everywhere and if the Universe is open this everywhere knows no bounds. The question is: how seriously do we take infinity? We can observe only a small part of the universe (see Figure 1.2). If we *do* take homogeneity, isotropy, and infinity seriously, we are claiming we know *exactly* how the infinite universe behaves *everywhere* by observing our one tiny little patch.

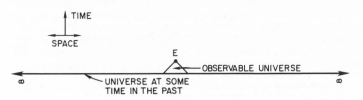

Figure 1.2. An infinite universe. Because nothing can travel faster than light, we can study only that part of the universe from which light has already reached us. In a relativistic spacetime diagram like the one here, light travels along 45° lines. If the earth is now located at point E, the observable universe is that area within the "past light cone" as shown. (The long horizontal line represents the state of the universe at some time in the past, say at the instant of the Big Bang.) But if the universe is truly infinite in extent, then there is an infinite amount we have not seen. Can we justify the assumption of homogeneity and isotropy by observations only within the past light cone?

This is worse than an ant climbing atop a grain of sand in the Sahara and claiming the entire world is made of similar grains.

The difficulties with infinity are further illustrated by an amusing example concoted by one of us (G.F.R.) and G. B. Brundit. We may assume the probability of life arising in our sector of the universe is nonzero.* For illustrative purposes, let's say the odds are 1% that life arises in our vicinity. If the rest of the universe is exactly the same, then the probability that life arises in any other sector is also 1%. Thus, if we take at least one hundred sectors, the probability of finding life approaches 100%. We are not done. We are assuming an *infinite* universe with an infinite number of sectors. Hence, there must be an *infinite* number of occurrences of life. Now, again assuming uniformity, some other life will be based on DNA, and DNA can produce only a *finite* number of configurations (say 10^{78}). Therefore, in an infinite universe there should be an infinite number of genetically identical species! In

* We could still be here if the probability of life arising were zero, but then we would be a unique occurrence and thus define a center of the universe. However, since we are assuming homogeneity and isotropy, we take it that the probability is nonzero everywhere.

fact, each one of us has not *one* genetically identical twin out there, but an infinite number of them. Do you believe this? No? Then we leave you to provide your own answer. We provide a few possibilities in the following sections.

3. SHAKE AND BAKE: ANISOTROPIC AND INHOMOGENEOUS MODELS

In the previous section, we mentioned that followers of the Principle of Greatest Probability find it difficult to believe the universe was created with exact isotropy and homogeneity. By 1968 this very conceptual objection led Charles Misner at the University of Maryland to propose an alternate scenario: the universe was created in chaos and later assumed the uniform, isotropic appearance we observe today. The question is, what form did chaos assume and what mechanisms brought about the isotropization of the universe?

The easiest approach is to relax the assumption of isotropy while maintaining homogeneity. (Exercise: describe to yourself anisotropic but homogeneous milk.) Thus, while the universe still looks the same at any location, its characteristics now change according to direction. A typical feature of anisotropic cosmologies is that they expand at different rates in different directions. As a not-very-accurate mental picture you might imagine an isotropic universe expanding like a sphere and visualize an anisotropic universe expanding like an ellipsoid (an American football) or even an object that keeps changing shape (technically an ellipsoid whose major and minor axes continually change direction).

Whereas in an isotropic universe the average flow of particles is the same in all directions—outward from the point of observation—in an anisotropic cosmology the various expansion rates cause particles to stream with many different velocities and hence energies that now depend on direction. These streaming particles will collide at early times and transfer energy among themselves until they all finally have the same energy and the universe is more isotropic. The so-called viscous damping of anisotropy is very much like what happens when you open a window from a warm house onto a cold day. The situation is anisotropic (and inhomogeneous): the cold particles outside have a lower energy than the hot particles inside. But they mix and transfer energy until the house is at the same temperature as its surroundings—isotropy. (We might mention that Misner called his original model "The Mixmaster Universe.")

Now such cosmic blending might seem very attractive to followers of the Principle of Greatest Probability, but anisotropic cosmologies have their own problems. First, they do not get rid of the singularity. To the contrary, they make it worse; if you imagine the universe to be collapsing as we go backward in time—instead of expanding as we go forward in time—a collapsing anisotropic universe forms a singularity faster than an isotropic universe. (For more details, see "On Cosmology" by L. C. Shepley and T. Rothman in *Frontiers of Modern Physics*.) Furthermore, there is a problem with element production. Beginning with the work of Kip Thorne in 1967, it has generally been thought that even a small amount of anisotropy *raises* helium from the amount produced in the standard model above the 25% observational limit. In the last few years, other investigators have claimed anisotropy *lowers* helium—drastically. The most detailed study of this question has recently been completed at the University of Texas by Richard Matzner and T.R., and the conclusion currently seems almost unavoidable that anisotropy indeed raises helium—and more sharply than previously thought. ("Currently," in physics, means about the lead time for publication—from 3 weeks to 6 months.) To keep helium below the 25% limit imposed by astronomers requires that the universe was highly isotropic at the time of element formation.

In addition, observations of the cosmic microwave background are continually becoming more refined. Recent work by Fixen, Cheng, Wilkinson, Lubin, Epstein, and Smoot, among others, shows that the microwave background is extraordinarily isotropic now. If one projects these observations back to the time of element creation, one reaches the same conclusions that we drew from the nucleosynthesis study just mentioned—the universe was *highly* isotropic at three minutes after zero.

At this point, the reader might justifiably feel the observations demand that anisotropic cosmologies be thrown out. Yet, there is a paradox involved. Although the universe appears to have been isotropic even at three minutes, it may be that we are compelled to use an anisotropic Big Bang to explain the existence of certain particles such as photons. There is no denying photons exist, and in great quantities too. There are approximately one billion photons in the universe for every proton. Now, according to quantum theory, particles may be created from the "vacuum." One should not think of this vacuum as empty, but as a storehouse of energy from which particles may arise. However, according to an argument originally put forth by Leonard Parker, particles like photons and neutrinos *cannot* be created from the vacuum in an isotropic universe.

Thus, if we believe that all particles originated from the vacuum, we may be forced to assume the universe started anisotropically. Well then, what happened? Reasoning along the lines of Misner, Ya. B. Zel'dovich suggested in 1970 that the particles created in the anisotropic universe collided with each other in the way described above for Misner's Mixmaster and quickly isotropized—all within 10^{-43} seconds!

This is a rather attractive idea: the universe begins anisotropically, photons and other particles are created, and the very creation of these particles causes the universe to become isotropic. While the idea is very reasonable and may be The Answer, we must confess that the actual calculations purporting to verify this phenomenon strike us as severely limited, if not inconsistent, and to claim the universe became isotropic within 10^{-43} seconds is a bit premature. Actually, we must confess that the new Inflationary scenario may bypass the whole problem completely. Inflation is briefly discussed in Section 5.

We have spoken at length about anisotropic cosmologies. What about inhomogeneous universes? Because the world may now have different properties at any point, inhomogeneous cosmologies are very difficult to study—few assumptions can be made— and less is known about their behavior. (See Figure 1.3 for a new result and a good example of inhomogeneity.) Nonetheless, inhomogeneous cosmologies have one great feature in their favor—the Principle of Greatest Probability. That is, unless some other principle overrides our conception of what is probable and what is not, we would expect the initial state of the universe to be inhomogeneous. What's more, inhomogeneous models have some intriguing differences from the standard case that are worth looking at. Particularly, since the universe is now unruly and inhomogeneous, the Big Bang could occur at *different* times in *different* places—the singularity could conceivably be going off *now* somewhere in the universe. Indeed, several Soviet physicists have sug-

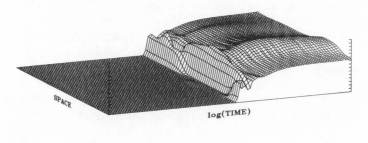

```
RUN     4
HE3
SCALE       0. 2000E-05
MAX HE3     0. 0000E+00        MAX  HE3     0. 2315E-04
MIN HE3     0. 0000E+00        MIN  HE3     0. 1843E-04
(INITIAL SLICE)                (FINAL SLICE)
```

Figure 1.3. The first numerical calculation of primordial nucleosynthesis in an inhomogeneous cosmological model by Joan Centrella, Richard Matzner, Jim Wilson, and T.R. The universe is divided into many cells along the "space" axis, and conditions vary from cell to cell (inhomogeneity). At time equal to three minutes on the "time" axis, helium formation begins. Here, we plot along the vertical axis the amount formed of an isotope called helium-3. You see the variation of helium from cell to cell as it is being synthesized. After several hundred seconds helium synthesis slows down and the graph flattens out. The helium-3 produced in this model varies from a minimum of 1.84×10^{-5} to 2.31×10^{-5}. In a homogeneous model there would be no ripples and no variation in helium.

gested that such delayed "Little Bangs" could be the power source for quasars. However, this proposal is not an essential feature of inhomogenous cosmologies—such models might very well be inhomogeneous at early times only.

This brings us to an interesting point. How can we reconcile an inhomogeneous model with the astronomical evidence that the universe is apparently homogeneous? There are several possibilities. First, as we said earlier, we cannot really verify that the universe is homogeneous because we cannot move very far from our position on earth. (Recall that the definition of homogeneous requires the universe to be the same *everywhere* we move.) What we do observe looking around us is that the universe is isotropic. Now, *if* it is the case that the large-scale universe is truly homogeneous, it is very possible that the same processes already described that damp out anisotropies also serve to smooth out inhomogeneities—and they may do so in a very short amount of time.

There is a second possibility. The universe may really be inhomogeneous on a large scale. Generally, if a universe is inhomogeneous everywhere it will appear anisotropic as well *unless* we are at a privileged position (Figure 1.1c). Hence, our universe might appear to be isotropic only because we are located in a special position in an otherwise inhomogeneous cosmos. You might argue that this harks back to pre-Copernican ways of thinking— and indeed it does. Nonetheless, such a possibility might be justified by the *anthropic principle.** In this instance the anthropic principle would state we can observe the universe only from positions favorable to life. This is a very attractive idea if it can be made good. It then becomes a major problem to explain why life should be preferentially evolved in a region where the universe appears isotropic. No one has yet succeeded in doing this convincingly, but the anthropic connection remains a fascinating realm for future work.

* There is much literature on the anthropic principle. By far the most complete account is John Barrow and Frank Tipler, *The Anthropic Cosmological Principle* (Oxford: Oxford University Press, 1986). Less technical is the article by Carr and Rothman in *Frontiers of Modern Physics*, as well as the article by Rothman in *Discover* (May 1987).

The third explanation for the apparent homogeneity of the universe is quite different and also alluring in many ways. It requires that we live in a small inhomogeneous universe that we have already seen around many times. To visualize this, imagine standing with a small group of people in a room with mirrors on all the walls. Looking around, you will then see images of thousands of people stretching away into the distance in all directions. Now consider instead a set of, say, one million galaxies in a large region of space, surrounded on all sides by mirrors. The appearance will be that of a nearly homogeneous, isotropic collection of many *more* millions of galaxies stretching into the distance. This is exactly the observational situation in a small universe that is closed on itself in space*: virtually whatever the distribution of galaxies and matter in the small, basic cell, it appears as if the observer is in a large universe that observationally resembles the standard, FLRW cosmology. Thus, this "funhouse" model provides a nice explanation for the apparent homogeneity of the universe—the galaxies we see in the distance look like those near us because they are indeed the same set of galaxies!

One might at first glance imagine this hypothesis would be easy to disprove—if we really lived in a hall of mirrors, we would see each galaxy (including our own) many times over. How is it we have not detected such a state of affairs? The answer is that detection would be far from trivial (see Figure 1.4). Because light travels from the galaxy to the observer along different paths, the various images from this galaxy would depict it at various times in its history and therefore at different states of its evolution. Each image might intrinsically be a different color or brightness, or perhaps even a different shape from the others. This situation would already make it difficult to recognize the various images as actually originating from the same object. The confusion is compounded by the fact that intergalactic and interstellar absorption might vary along the different paths of observation as well, chang-

* Super technical point: the reader may wonder how a small universe can contain enough matter to be closed in the sense we defined "closed" in Section 2. A universe may be closed in another sense—topologically (say, like a doughnut).

ing the image further. Moreover, the images of this galaxy will all have different redshifts, usually a key indicator of distance. Clearly, in practice, it is very difficult to tell if we have seen the same galaxy, or even clusters of galaxies, when looking in different directions. So, in our hall of mirrors we have an intriguing and attractive way to reconcile the Principle of Greatest Probability (and hence inhomogeneity) with observational similarity to the standard model. It is a possibility that needs further investigation.

To conclude this section, we emphasize that we are not at present able to determine whether any of these inhomogeneous models is a better model of the real universe than the FLRW. However, we are able to maintain that these are at least conceivable alternatives that need investigation, and that a scientific approach to cosmology must be based on a consideration of all the alternatives, not just the one family of models that happens to be the simplest to consider.

4. SCIENCE OR NUMEROLOGY?
VARIABLE-G THEORIES

In the previous section we talked about alternatives to the FLRW universe which changed the *geometry* of the spacetime. We asked what would happen if an isotropic universe became anisotropic inhomogeneous and we discussed the results. Of course, nothing prevents us from leaving the geometry unaltered and tinkering with other aspects of the universe. One popular alternative is to allow the fundamental constants of nature to change. Not surprising many physicists regard the values of the natural constants to God-given and refuse to take seriously any theory in which vary. Nonetheless, one should be open-minded about such th we do not yet have any theory which predicts the values o constants of nature (see, however, ''Coincidences in Nature by Carr and Rothman in *Frontiers*), and there is no a prior son—except the Principle of Simplicity—to assume they ar stant. For instance, in the laboratory we *measure* the value gravitational constant G to be roughly 10^{-7} cgs units, but w no idea why this should be so. We also have no a pri

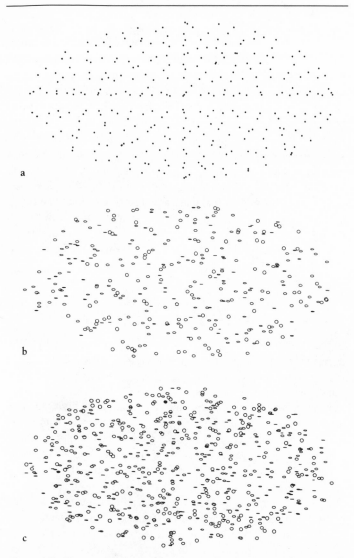

Figure 1.4. Computer simulations of a small universe model by Gunhild Schreiber. The universe is similar to the "funhouse" model described in the text, but, instead of being surrounded on all sides by mirrors, the universe is closed in the manner of a doughnut (torus) which we have

son to claim G always had this value. On the other hand, is there any reason to suspect the value of G has changed over the eons? Indeed, there are two distinct philosophical arguments which motivate "variable-G" theories. Unfortunately, the two arguments get jumbled up even in the scientific literature; we will try to keep them straight here.

The first motivation comes from the famous "Large Number Hypothesis" of Dirac, first put forth in 1938. It had long been noted that the ratio of the electrostatic force between an electron and a proton to the gravitational force between the same two particles was approximately equal to the ratio of the size of the universe to the size of an electron. It is easiest to write this coincidence in symbols so, if you forgive us,

$$\frac{\text{Force}_{\text{elec}}}{\text{Force}_{\text{grav}}} \sim \frac{\text{Size}_u}{\text{Size}_e}$$

or

$$\frac{e^2}{Gm_em_p} \sim \frac{m_ec^3t_u}{e^2} \sim 10^{39}. \tag{1}$$

In these expressions, $\sim$ means "approximately equal"; the size of the universe $= ct_u$, where $c =$ speed of light, $t_u =$ age of uni-

---

seen as round many times. In (a) the universe contains only one galaxy but 312 images. Due to the toroidal nature of the model, distinct patterns are visible which do not occur in the real universe. Hence this model is ruled out. In (b) there are 50 galaxies, 328 images, and the patterns are less distinct. In (c) there are only 30 galaxies in the basic cell but we are seeing more times around the torus and 612 images are produced. In (c) the patterns are still noticeable but less distinct than in (b). How many galaxies are needed in the basic cell before the model becomes indistinguishable from the observed universe is still an open question. (The number of images depends not only on the number of actual galaxies but on their position and cutoff redshift as well—how many times we've seen round. The cutoff redshift in [a] is 2.00, which explains the relatively large number of images for only one galaxy. In [b] the cutoff is the much smaller value 0.2, and in [c] it is 0.5.)

verse; the size of an electron $= e^2/m_e c^2$, with $e =$ charge of electron, $m_e =$ mass of electron; $m_p =$ mass of proton; $G =$ gravitational constant.

Now, the reader who bothers to work out the units in expression (1) will find they have all canceled out, leaving the two ratios "dimensionless." (They had better cancel; a force over a force or a size over a size has no units.) The result is rather startling because in physics most naturally arising dimensionless numbers are of order one, yet here we have two, both of an almost inconceivable size, 10^{39}, and both approximately *equal*. The other thing to notice is that, except for the age of the universe t_u, all the other quantities are fundamental constants. Why should the age and hence size of the universe have anything to do with the size of an electron? Maybe nothing. The approximate equality (1) could be just a *coincidence* and we are at liberty to dismiss it as such. Dirac took it seriously and suggested that for mysterious reasons this relationship always holds true. If so, we immediately have that G is proportional to $1/t_u$, or $G \propto t_u^{-1}$. In other words, *the value of G decreases as the universe gets older*. The reader is objecting: why should we not hold G constant and let m_e, m_p, or e vary? One could, but notice that m_e, m_p, and e are *atomic* constants, that is, they govern phenomena in atoms, whereas G governs gravitational phenomena—phenomena on the scale of planets and galaxies. In order not to create an entirely new atomic theory, Dirac chose G to be the variable.

Let us proceed. The total number of protons in the observable universe—call it N—is found* to be about 10^{78}. Note that 10^{78} is just the square of 10^{39}. Another coincidence? Perhaps. But again Dirac assumed it to be true. Using (1) immediately gives

$$N \sim \left(\frac{e^2}{Gm_e m_p} \right)^2 . \qquad (2)$$

---

* The number of protons in the universe N is just $N = nV$, where n is the number of protons per cubic centimeter and V is the volume of the observable universe. Since $V = (4/3)\pi R^3$, where $R = ct_u \sim 10^{28}$ cm and n is known from observations to be roughly 10^{-6} cm^{-3}, we get $N \sim 10^{78}$.

Since, G is proportional to $1/t_u$, expression (2) then tells us that the number of protons is proportional to t_u^2 [$N \propto (1/G)^2 \propto t_u^2$]. This means that the number of protons in the universe is increasing with time; in the Dirac theory *matter is created*.*

Now, it turns out that the Dirac theory is just one of a class of theories termed "scale-covariant" theories which have been extensively developed, most notably by Victor Canuto and his collaborators. In some of these theories matter is created, in others not. In addition, one is free to specify how G varies with time, that is, the relationship need not be $G \propto t_u^{-1}$.

While mysterious cosmic coincidences are fascinating and consequently many theories are based on them, we must stress that there is absolutely *no* solid evidence that indicates G varies with time. What's more, you would expect a change in G to affect the size of planetary orbits and anything else where G appears in the equations. One must be careful here because, it turns out, the effect of a variable G in scale-covariant theories is often more subtle than one would naively expect. Nonetheless, the naive Dirac theory is apparently already ruled out by observations of the orbital periods of the planets. Tests on other scale-covariant theories are difficult to make because one does not know how G varies with time and there are many possibilities. But, for instance, the amount of helium formed during primordial nucleosynthesis depends on how fast the universe expands; this in turn depends critically on the value of G at that time. It is difficult to believe that a larger value of G in the past would produce values of helium consistent with observation. Actually, T.R. has investigated this problem for at least one choice of variable-G theory and found that at three minutes after the Big Bang, G must have been essentially what it is now.

We mentioned there is a second motivation for variable-G theories. This is Mach's Principle, which states that absolute motion does not exist and only motion relative to the bulk of the matter in

---

* The sophisticated reader may know that in the FLRW model N increases simply because more particles are coming into our horizon as the universe expands. But this dependence is only $N \propto t_u$.

the universe is meaningful. For example, inertial forces, such as the centrifugal force we feel when a car rounds a corner, would not exist if the fixed stars were absent. Using Mach's Principle one can show in a crude way that the following relationship should hold:

$$GM \sim Rc^2, \tag{3}$$

where again G is the gravitational constant, M is the mass of the observable universe, and R is the radius of the universe. Now, relationship (3) is also seen to be approximately true in the real universe. It has often been called a cosmological coincidence like (1) and (2), but it is firstly a prediction and secondly involves two cosmological quantities (M,R), whereas the other coincidences each involved only one. Using $R = ct_u$, relationship (3) can be rewritten as $G \sim c^3 t_u/M$. If we believe G is a constant and assume c is as well, then this says that t_u/M must be fixed by the theory. It is not obvious that the age of the universe divided by its mass should be a constant. The only alternative is that G must vary with time.

This argument was first given in 1953 by Dennis Sciama as the basis for a variable-G theory. Similar reasoning in turn led to the famous Brans-Dicke theory of 1961, which also has a variable G. Unfortunately, the Brans-Dicke theory has been constrained so much by observations of the binary pulsar's oribital period that it is virtually the same as standard relativity. Probably the last believer in the Brans-Dicke theory died at the Port Authority Bus Terminal five years ago. Thus, as the sun slowly sets in the west, we bid a fond farewell to variable-G theories and turn to more profitable pursuits. Onward.

5. SINGULARITIES AND POTPOURRI

We have frequently said the singularity is the Big Stop, the conceptual cataclysm, the ultimate barrier. This is true. To avoid the initial catastrophe is the primary reason many, if not most, alternative cosmological models have been developed. Once again, the best way to think about this problem is to imagine the universe to be running backwards and ask what possible mechanisms might prevent it from collapsing into one mathematically sized point.

Once upon a time it was hoped that anisotropic models would prove to be the solution—since particles are streaming in all directions, they might somehow "miss" the singularity. But as stressed in Section 3, anisotropic singularities contain *worse* singularities than the standard model.

Another possibility would be to come up with a source of negative pressure that would halt the universe from collapsing into a singularity and instead cause it to "bounce." Believe it or not, there are several ways of doing this. The first is to reinvoke the cosmological constant. Recall we discussed how Einstein added such a constant to his equations to make his universe static. This constant could be thought of as containing a repulsive force just strong enough to balance the universe between expansion and collapse. Well, if you made the constant larger and more repulsive, it might keep the universe from collapsing into a singularity.

The good news is that you can indeed make the cosmological constant large enough to "bounce" the universe. The bad news is the cosmic microwave background. Astronomers are virtually certain this radiation was produced about 100,000 years after the Big Bang when the universe was roughly one thousand times smaller than it is today and at a temperature of about $3,000°K$. Now, if the cosmological constant caused a bounce, say from a previous collapse, this bounce had to occur at an *earlier* time than when the microwave background was created. If not, we would simply have no background radiation. To accomplish this, the constant would have to be *so* repulsive that we would currently see the universe *accelerating* rapidly outward, whereas if anything, we observe exactly the opposite—the expansion of the universe is *slowing down* due to the gravitational attraction of all the matter within it. The problem is even more serious when one considers primordial nucleosynthesis. If the supposed bounce took place after the three miniute mark of element formation (meaning the universe never got hot enough for nucleosynthesis), then we must explain the origin of the elements in some other way. "Cold Big Bang" theories do exist and try to produce helium and deuterium from primordial stars which came into existence not long after the Big Bang, but these theories are not very convincing and certainly violate the Principle of Simplicity. Returning to the cosmological constant, if the assumed bounce took place *before* nucleosynthesis times—allowing element formation in the standard way—the constant would have to be 10^{24} times larger than in the microwave background case and we would observe all the galaxies madly accelerating away from each other.

Large cosmological constants may be passé, but there is at least one recently famous theory that acts as if it has such a constant or negative pressure. This is the so-called "inflationary" scenario first proposed in 1980 by Alan Guth and in a more recent form by Andy Albrecht, Paul Steinhardt, and A. Linde. In Guth's theory the huge negative pressure drives the expansion of the universe exponentially fast for a brief period of time at around 10^{-35} seconds and this solves a number of problems in the standard model which have puzzled physicists for a long time. For instance, why is the observed matter density of the universe so close to the critial density needed to close it? Why is the ratio of photons to protons the large number 10^9? We do not have space here to give an adequate treatment of inflation and refer the reader to the article by Guth and Steinhardt in the May 1984 issue of *Scientific American*. We only remark that, first, many questions about inflation itself currently remain unanswered; and, second, in many ways, the model is just a return to the Steady-State theory introduced in 1948 by Bondi, Gold, and Hoyle.

The Steady-State theory was based on the "Perfect Cosmological Principle" which states that the universe should be homogeneous and isotropic not only in space but in time as well. In other words, the universe must have looked the same in the past as it does now. But since the universe *is* expanding and galaxies *are* getting farther apart from one another, the gaps must be filled in to ensure that the universe *now* looks the same as it did *then*. Using words we have uttered before, matter is created.

How has Steady-State managed this sleight of hand? Well, with a cosmological constant. Always remember that the cosmological constant can be regarded as an extra energy term in the Einstein equations (resulting in the negative pressure). Specifically, it is a constant energy per unit volume. So, as the universe expands, its volume increases and therefore its total energy does too.* Since matter and energy are really the same thing, by introducing a cosmological constant we must create matter.

Unfortunately, the Steady-State theory finds it virtually impossible to explain either the light elements or the cosmic microwave background, both of which require the universe to have been much different in the past than it is today, namely very hot. For this reason, all but the most rabid fanatics gave up the Steady-State theory around 1965 with the discovery of the microwave background.

A final note on Steady-State and Inflation. We do not want to leave the reader with the impression that the two theories are identical. They aren't. The inflationary epoch supposedly ended at about 10^{-30} seconds while Steady-State is supposedly good now. But both require an effective cosmological constant, both have matter creation, and both describe the expansion of the universe in the same way. In the Steady-State theory there is no initial cosmological singularity—by assumption, since the universe always looked the same as it does now—and the cosmological constant is truly constant. The inflationary scenario does not directly address the issue of the singularity, and, moreover, the cosmological constant, though once very large, is now zero.

So what do we finally do about the singularity? Inflation doesn't solve the problem, and Steady-State conflicts with observational evidence as do other Big Bang theories with large cosmological constants. There may be a way out. One might simply assume the existence of matter with negative energy—in other words, matter that is gravitationally repulsive. This in turn would produce the large negative forces needed to bounce the universe instead of letting it collapse into a singularity. At first glance such an idea might seem absurd, as all matter we observe in the universe is gravitationally attractive. But what is absurd at first glance may not be at

---

* Very subtle exercise: if the universe is infinite, how can increasing the volume increase the total energy?

second. It is thought that when the universe was very young, less than 10^{-43} seconds old, and very hot, $10^{32}°K$, quantum effects would come into play which would effectively cause matter to have negative energy. If this is actually the case, then it is quite likely a singularity can be avoided. But here we are standing on the edge of a deep abyss of speculation because a coherent theory of quantum gravity has yet to be developed.

We have presented the reader with a number of cosmological models: isotropic, closed, open, anisotropic inhomogeneous, halls or mirrors, Variable-G, with constants, without constants, Big Bang, and Steady-State, Inflation, and Quantum. Each has its attractive features and each has its failings; undoubtedly the reader has a headache. In the end we must simply say we don't know. But before the end, we should remind our audience that the Garden of Cosmological Delights is very large. Someday you may run into tired light (the universe is not expanding at all—light gets "tired" as it travels and this causes the redshifts astronomers observe), or Kaluza-Klein (the universe has eleven dimensions). And what about other supergravity theories (the universe has any number of dimensions)? Then there are also early dustlike stages and scalar fields, GUTS and Susy GUTS. . . . The list could go on almost indefinitely. But for the time being, we have run into a singularity and will call a very abrupt halt.

2·METAFLATION?

(WITH G. F. R. ELLIS)

1. ETYMOLOGY RECAPITULATES ONTOLOGY

Physics and metaphysics. Nature and beyond nature. Or so the terms are ordinarily used. Physics is a science; it deals with questions that can be decided by observation and experiment. Metaphysics deals with issues that do not admit defeat so easily: the existence of God, the content of the soul. In the twentieth century, scientists pride themselves on maintaining the distinction between them. How the Big Bang occurred is a question cosmologists attempt to answer: *why* the Big Bang occurred is not.

As we have just used them, the terms physics and metaphysics might suggest science versus religion. We do not want to give this impression. In fact, the word "physics" derives from the Greek *physika*, which literally means "things of nature." "Metaphysics" comes from *meta ta physika*, literally "after the things of nature." As far as we can tell, the term was first used by Aristotle's Hellenistic editors to refer to the texts that followed his books on physics. Medieval theologians convinced themselves that these texts came "after the things of nature" because the subjects dealt with were farther removed from direct perception than those of the *Physics*. Since then, "metaphysics" has always connoted something that transcends physics and, since the time of Kant, has taken on its common meaning—the study of questions that cannot be proven or disproven by scientific experiments. We will continue to use the term in this way, although you may prefer

29

H. L. Mencken's splenetic characterization: "Metaphysics is almost always an attempt to prove the incredible by an appeal to the unintelligible."

In the past, scientists would not have agreed with Mencken. Newton had no difficulty, on the one hand, in postulating that the force of gravity obeys an inverse-square law and, on the other hand, declaring that God was necessary to prevent the universe from collapsing under the same law of gravity. The first is a testable hypothesis, the second is not. Many scientists of the seventeenth and eighteenth centuries, including the great Christiaan Huygens, believes the cosmos was filled with inhabited worlds because such an account of the universe was more worthy of the Infinite Creator than a universe occupied by one lonely Earth.

Nevertheless, Newton's mechanics and its descendants proved so successful at eliminating God and similar untestable hypotheses from scientific explanation that, three hundred years later, science has a different voice altogether. Steven Weinberg writes in *The First Three Minutes* (Basic Books, 1977): "The more the universe seems comprehensible, the more it also seems pointless."

In this essay we are going to discuss a cosmological theory that you have probably already heard about—inflation—but we are going to discuss it with the following questions in mind: Is the twentieth-century distinction between physics and metaphysics still valid? Has cosmology returned to the theological stage in which it makes untestable predictions? Are the assumptions on which cosmology rests verifiable? Should present-day cosmology be called cosmology or metacosmology?

2. CONUNDRUMS

In "The Garden of Cosmological Delights" we spent some time discussing the standard Big Bang model and its successes. For our present purposes we need recap only a few details. The standard Big Bang model (or FLRW model in honor of Friedmann, Lemaître, Robertson, and Walker) assumes that the universe started to expand at some finite time in the past, thought to be between ten and twenty billion years ago. The FLRW world is also homogeneous and isotropic. *Isotropic* means that the universe appears the same to us in all directions. *Homogeneous* means that if we traveled far from Earth to any point in space, the universe would appear the same there as it does in our neighborhood. *A homogeneous and isotropic universe contains no irregularities that could distinguish one location from another. The matter is taken to be distributed in an absolutely uniform fashion everywhere.* For this reason a simpler universe than the standard Big Bang model is difficult to imagine. It is important to note that the real universe is observed to be highly isotropic when we look at it on a large enough scale, but we cannot verify that the universe is homogeneous because, by cosmological standards, we cannot travel very far from Earth.

If the density of matter in the FLRW model exceeds the so-called "critical density" of about 2×10^{-29} grams per cubic centimeter, the universe will eventually halt its expansion and recollapse. In this case the model is usually called "closed." If the matter density is less than the critical value, the universe will go on expanding forever. Such a model is usually referred to as "open." If the matter density is exactly critical, the universe is termed "marginally bound" or "flat." The two most famous triumphs of the standard Big Bang are its predictions of the cosmologically observed abundances of helium and deuterium and of the existence of the cosmic microwave background radiation.

We have also, in "The Garden of Cosmological Delights," mentioned several of the failures of the standard model. The most serious physical difficulty is the existence of a singularity at the instant of the Big Bang, that is, a time when the pressure, density, and temperature all become infinite and the known laws of physics break down completely. Spacetime, and indeed physics itself, comes into existence at this time. A serious philosophical difficulty is to understand why the universe should have started off exactly isotropic and hence exactly uniform everywhere, which is what the standard model assumes.

As it turns out, the standard model has other problems, which are not so popularly known. It is these conundrums that Alan Guth set out to solve in his "inflationary universe" scenario in his celebrated *Physical Review* paper of January 1981. In doing so, Guth created a bit of inflation himself, an industry which has generated hundreds of papers and has not ceased production at the present moment.

The first of the puzzles that Guth addressed has been emphasized during the last decade by Robert Dicke and James Peebles of Princeton University and is now known as the "flatness problem." The origin of the flatness problem lies in the fact that astronomers measure the density of matter in the real universe to be within roughly a factor of ten of the critical density. We might make an even stronger statement: most, if not all, observational and theoretical studies indicate that the baryonic matter density (the density of neutrons and protons) lies between about 4% and 10% of the critical value. In principle, the matter density could be a billion times less or a billion times more than critical. Other than the anthropic principle there is no fundamental reason why this should not be so. But we happen to observe the real universe to lie within a factor of ten of being flat.

This could be a coincidence, but physicists feel uncomfortable with coincidences and search for underlying causes. More to the point, it is not difficult to show that if the universe is currently within a factor of ten of being flat, then at one second after the Big Bang, the start of element formation, it must have been flat to approximately one part in 10^{15}. And at 10^{-35} seconds, the famous Grand Unified epoch, the universe must have been flat to one part in 10^{50}. Such fine-tuning strikes many physicists as in need of an explanation. Other physicists point out that, in the standard model, the universe is exactly flat at the instant of the Big Bang, so it's not surprising that it's tuned to one part in 10^{50} at 10^{-35} seconds. For skeptics there are several alternative ways to phrase the flatness problem. We give one of these now, though you will have to accept our word that it is a statement of the same problem.

We estimate there to be about 10^{87} photons in the observable universe. This might seem as reasonable a figure as any other, but there are two aspects of the number which, taken together, trouble some cosmologists. First, it is a very large number by ordinary standards and, second, it is a number without any units like centimeters or grams. In physics, most naturally occurring "dimensionless" numbers are of order one. Thinking naively you might expect one photon in the universe instead of 10^{87}. Admittedly, it would be a very dark universe but that is another story. You might find this argument reminiscent of that connected with the Dirac Large Numbers in our previous essay, "Garden." There, we asked for an explanation of the number of baryons in the observable universe, which is roughly 10^{78}. This number is not unrelated to 10^{87}; to the contrary, the relationship is explained by the new Grand Unified Theories (GUTS). Thus, to ask for an explanation of the flatness problem is essentially to ask for an explanation of the Dirac Large Numbers. Guth's solution, on the other hand, is very different from Dirac's. Nevertheless, if you are inclined to believe the Dirac Large Number questions are a pseudoproblem, you are for the time being entitled to dismiss the flatness problem in the same breath. In keeping with our theme of physics and metaphysics, here is the first major point: the flatness problem is a metaphysical and not a physical dilemma. Obviously, the reason for 10^{87} photons in the universe is that God said, "Let there be light."

A second conundrum, originally clarified by Wolfgang Rindler as far back as 1956, may strike you as more substantial. It is known as the "horizon problem." To explain the horizon problem we must cautiously reveal a fact about the standard model that most popular accounts try to avoid: there are two sizes to the universe. The first of these is simply the age of the universe (t) multiplied by the speed of light (c). We will denote this distance as $h = ct$. Because no signal can travel faster than the speed of light, you cannot in a time t receive information from any distance farther than h. A better way of putting this is: you can have no knowledge of events taking place beyond your horizon. The distance h is aptly called the horizon distance. The observable universe lies within our present horizon.

The second important measure of distance in the standard model is usually called the cosmic scale factor and is denoted by R. In the case of a closed model, R can properly be identified with the radius of the universe. In an open model this identification is not strictly correct (since the spatial size of the model is infinite), so we will restrict our attention to closed models for conceptual simplicity. A closed model may be likened to a balloon of radius R that can expand or contract. If the balloon expands, galaxies or particles painted on the surface will get farther apart from one another, but the total number of galaxies or particles does not change. Let us suppose there are 10^{87} photons in the real universe within the radius R. This number does not change under expansion or contraction. If R decreases by a factor of two, the average separation between photons also decreases by a factor of two, and the 10^{87} photons are now contained in a volume that is smaller than the original by a factor of eight (since volume goes as R^3). There are always 10^{87} photons in the universe, regardless of size.

Now, it turns out that, as the universe expands, R increases more slowly than h (see Figure 2.1 for more details). Conversely, as we go back in time, h is decreasing *faster* than R. Assume that today all 10^{87} photons in the universe can be observed. This means that the horizon size equals the scale factor ($h = R$). But as we project backwards, h is shrinking with respect to R. Since only those photons within the horizon are observable, as we go backwards in time, we see a smaller and smaller fraction of the 10^{87} photons in the universe. A straightforward calculation shows that at one second after the Big Bang, h was only about $10^{-9}R$ and only 10^{60} photons could be observed, an insignificant fraction of 10^{87}. At 10^{-35} seconds, h was about $10^{-27}R$ and roughly one million photons could be found within the horizon. At 10^{-43} seconds, essentially no particles were within the horizon.

This result has a simple but surprising interpretation. It says that as we go toward the Big Bang, eventually the entire universe—and all the particles in it—lie outside the horizon distance h. Because one particle cannot exchange information with any other particle outside its horizon, we see that in the early universe virtually all particles were highly incommunicado. At 10^{-35} seconds, a given photon could interact only with its nearest 10^6 neighbors and had no knowledge of what the other 10^{87} photons in the universe were doing. Should this worry us?

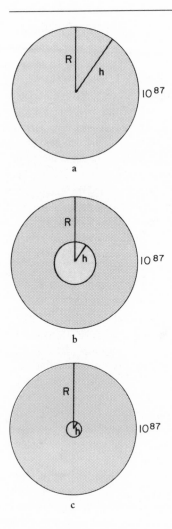

Figure 2.1. The horizon problem. A given particle can only observe and interact with other particles that lie within its horizon, h. Suppose there are 10^{87} photons within the radius of the universe R. If today $h = R$, then we can observe all 10^{87} photons in the universe. This situation is shown in (a). But for the first 100,000 years after the Big Bang, h goes as t, R as $t^{1/2}$, and thus h/R as $t^{1/2}$. The horizon is approaching zero faster than R. At one second after the Big Bang (b), h/R was about 10^{-9} and the volume of the observable universe was only a fraction $h^3/R^3 \sim 10^{-27}$ of the total universe. Consequently, only $10^{-27} \times 10^{87} = 10^{60}$ photons were within a given particle's horizon and could interact with it. At $t = 10^{-35}$ seconds (c), h/R was about 10^{-27}, the horizon volume was the minute fraction 10^{-81} of the total, and only 10^6 photons were in causal contact. At 10^{-43} seconds (not shown) no particles were within the horizon, but the entire concept of time may break down here—the epoch of quantum gravity. We know the universe was smooth by about one second after the Big Bang. How did this come about if only an insignificant fraction of the particles could interact?

As we have stressed, the universe is observed to be highly isotropic. Many cosmologists find it extremely improbable that the universe started off in an absolutely isotropic fashion. To them it seems far more likely that the Big Bang occurred in an absolutely chaotic fashion and gradually isotropized, or became regular. Cosmologists who are inclined toward this point of view may be termed advocates of the Principle of Greatest Probability. Unfortunately, the horizon problem puts members of the Greatest Probability Sect into a serious blind alley because the universe could not then have become isotropic. Since particles farther apart than the horizon distance h could not interact after the Big Bang, there is in principle no method by which they could have smoothed out any irregularities on scales larger than this same distance. The question posed by the horizon problem can be rephrased: why is our present universe not chaotic, like the world in general and government bureaucracies in particular?

One answer was provided by the Soviet astrophysicist Ya. B. Zel'dovich who terminated a similar discussion with T.R. by exclaiming, "I believe the universe started isotropically!" Zel'dovich, then, might be termed a follower of the Principle of Simplicity, which holds that the universe began in the simplest possible manner—with complete isotropy. If you are inclined toward this point of view, there is no need to explain the isotropy of the universe—you have assumed it from the beginning. The horizon problem becomes irrelevant; true, particles near the Big Bang were unable to communicate and thereby isotropize the universe, but any such contact would have been superfluous because the universe was already expanding isotropically.

We conclude that the second problem inflation sets out to solve may also be metaphysical. God said: "Let the universe be isotropic." Both the present authors belong to the Sect of Greatest Probability, which means that the horizon problem is a serious dilemma. However, we claim that the horizon problem and flatness problems are not independent. On the one hand, this makes it harder to dismiss the flatness problem as a pseudoproblem; on the other hand, inflation is not tackling two distinct puzzles. God said, "Subtlety but not malice."

3. INFLATION

It is not necessary for the theme of this essay to explore all the alleyways of Guth's original "old" inflation, nor the "new" inflation proposed in 1982–83 by A. Linde of the Soviet Academy of Sciences and independently by Albrecht and Steinhardt at the University of Pennsylvania; nor the even more recent "chaotic" inflation, "supersymmetric" inflation, and countless variations thereof. Nevertheless, a basic understanding of inflation will be useful to what we have to say later.

Imagine yourself to be an ant crawling on the surface of the balloon described earlier. In its uninflated state, the balloon may be so small that the curvature of the surface is highly visible and you are able to completely circumnavigate the globe in a short amount of time. Suppose the balloon is suddenly inflated thirty or forty orders of magnitude, that is, by a factor of 10^{30}, 10^{40}, or even 10^{100}. As a lowly ant afoot, the surface will appear to you flatter than the steppes of Asia, and the horizon, which was previously very close, will now be so distant that you cannot see it.

We have not idly chosen the balloon analogy. Guth's inflationary scenario proposes that much the same occurred in the real universe. At roughly 10^{-35} seconds after the Big Bang, the universe underwent a brief period of exponential expansion until, say, 10^{-30} seconds, which inflated the primordial fireball by at least thirty orders of magnitude. We will sketch how this works in a moment, but for now the important point is to realize that such an inflation would make the universe extraordinarily flat. We have said that a flat universe is one at the critical density, so it should seem plausible to you that the inflationary period drives the matter density of the universe toward the critical value. In other words, regardless of the value of the density before inflation, after inflation the density is, to a high approximation, critical. The universe is, for all intents and purposes, flat, and the flatness problem disappears. That the density of the universe is now critical is the most important prediction of the inflationary scenario and it is essential to keep it in mind.

The horizon problem vanishes at the same stroke (which is not surprising if the two puzzles are not distinct). Recall that at 10^{-35} seconds, the horizon size was approximately 10^{27} times smaller than the scale factor R, so that particles could not interact. If the horizon were 10^{27} times larger or, conversely, if the scale factor were 10^{27} times smaller than previously thought, the horizon problem would disappear. As shown in Figure 2.2, inflation allows this possibility; at early times the scale factor was, contrary to the standard model, *smaller* than the horizon size, so particles could communicate and smooth out any irregularities.

It should be emphasized that inflation provides a necessary but not sufficient "isotropizing agent." Particles now have the opportunity to interact, but inflation does not specify *how* they interacted. One must either invoke some other mechanism to explain how the universe isotropized (such as the quantum damping of anisotropy discussed in "Garden"), or the inflationary scenario should itself show that, regardless of the initial state of the universe, the final result is always homogeneous and isotropic. We will return to this important point later.

We have outlined what inflation does. We now attempt to give an idea of why inflation takes place. Again, we must resort to an analogy. It is well known that if you take a flash of steam at a very high temperature and gradually cool it down, you can cool it far below 100°C and yet not see it condense. Steam that exists below the usual condensation point of 100°C is referred to as supercooled. The state is highly unstable, and if the smallest water droplet is introduced into the flask, the steam will quickly condense around it until the entire contents of the flask have turned into liquid water. With his original inflationary scenario, Guth proposed that the universe went through a similar phase transition. The universe expanded and its contents—photons, electrons, neutrinos, and other particles—cooled. The vacuum energy of spacetime cooled as well. In modern physics the vacuum of spacetime is far from empty. Instead, it is an arena where particles may be created and destroyed and which has many properties that can be calculated and measured. This "ground state" of the universe has

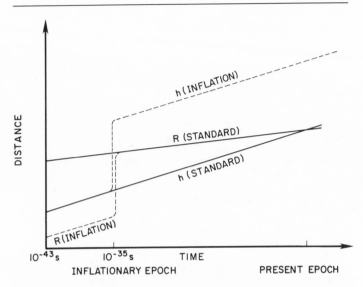

Figure 2.2. Inflation's solution to the horizon problem (highly schematic). The cosmic scale factor R and the horizon size h for the standard model are shown by solid lines. R and h are set equal at the present time as in figure 2.1. At 10^{-35} seconds, h/R was about 10^{-27} and only 10^6 photons were able to interact. This problem goes away if at that time R were smaller by the factor of 10^{27}. Inflation (dashed lines) shows how this could be accomplished. Contrary to the standard model, R starts off smaller than h by the required factor, but at 10^{-35} seconds both R and h were inflated by at least 27 orders of magnitude, as shown in the diagram. After inflation, R and h continue to evolve as in the standard model but the horizon is pushed far beyond the presently observed universe.

an energy density associated with it that also changes as the universe expands. In some ways the vacuum energy is similar to that associated with ordinary particles. Now, at about 10^{-35} seconds, the ordinary particles continued to cool, but the vacuum energy got "hung up" in a supercooled or metastable state (Figure 2.3). According to relativity, the expansion rate of the universe depends on the energy density of its contents. In the standard model, the energy density of the particles decreases as the universe expands,

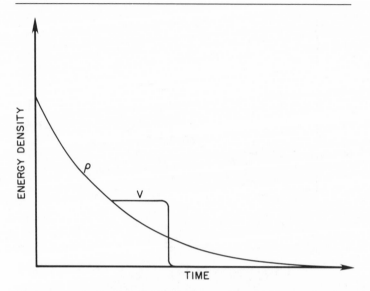

Figure 2.3. The expansion rate of the universe depends on its energy density (ρ). In the standard model, ρ decreases as shown and the expansion rate of the universe therefore goes down. Inflation proposes that at 10^{-35} seconds, the vacuum energy density (V) enters a metastable state and no longer decreases. The energy density of the universe is now much higher than in the standard model and the expansion therefore proceeds more rapidly as well. At some point, the vacuum energy decays to its present value and inflation ends. One difficulty with this scenario is that no one knows what the final value of the vacuum energy should be (see Section 4).

and so does the expansion rate. But the inflationary universe suggests that the vacuum energy got "hung up" at a higher, constant value than would be the case in the standard model. The expansion thus proceeds at a higher, constant rate than it would have in the standard Big Bang. This is inflation. If the vacuum energy remains constant for a long time, the universe can expand the required factor of 10^{30} or 10^{40}. At some point, condensation sets in; a "water droplet" is introduced into the supercooled system, the vacuum energy quickly decays down to the value it would have had under ordinary circumstances, and inflation ends.

We should mention that Guth did not do elaborate calculations to prove that the supercooled state of the universe *would* take place. He suggested that it was *plausible* for it to take place and *if* it took place, inflation would occur.

We should also point out that there is a severe problem with the original inflationary scenario. Guth realized that the decay of the vacuum energy would, like the condensation of the steam, occur at different times in different places. But unlike the flash of steam, the universe is expanding and the decay rate of the vacuum must catch up with the expansion of the universe for the condensation to be completed everywhere. Guth found that this did not generally happen and that the resulting universe would be highly irregular—some parts in one state and some parts in another. Since we observe the universe to be highly regular, this difficulty ruled out "old original" inflation.

The introduction of "new" inflation by Linde and Albrecht and Steinhardt solved this problem of the "graceful exit." The mechanism behind new inflation is conceptually more difficult than old inflation and we will not go into the details. (The reader may want to consult the article by Guth and Steinhardt in the May 1984 *Scientific American*.) It is enough to say that the vacuum energy exists in this scenario as well, and remains at a sufficiently high value for a sufficiently long time to drive inflation much as in the original model. The essential difference is that, in new inflation, one small bubble of space is inflated to encompass the entire known universe—and more. We do not see irregularities because they have been pushed far beyond our horizon.

New inflation had an additional advantage over old inflation in that it solved a third cosmological conundrum. This is the monopole problem. Grand Unified Theories predict that the universe should be filled with magnetic monopoles. A magnetic monopole may be thought of as an isolated magnetic charge, like a single north or south pole of a magnet. No one has ever seen such a thing. Magnets always come with both a north and a south pole. New inflation solves this problem in a trivial fashion. It says that the high density of monopoles in the early universe is simply diluted by the period of inflation. While in the standard model you would expect to find monopoles everywhere, inflation increases the volume of space to such an extent that you would expect, on average, one monopole in the observable universe, and we haven't found it yet.

Unfortunately, a number of diseases have been diagnosed in the new inflationary scenario, of which we now mention two. First, the universe apparently decays into the wrong vacuum. Loosely speaking, inflation puts us into a universe where the physics of the "ground state" is not the physics that the currently favored theory predicts. This vacuum will then decay into ours, but the decay occurs unevenly as in old inflation, leaving a highly irregular universe. Thus nothing has been gained.

Second, inflation predicts that certain fluctuations in the density of matter *will* exist after inflation comes to an end. It also predicts the size of these fluctuations. Now, other theorists believe they know how galaxies form. Galaxies are obviously fluctuations in the density of matter in the universe: the matter density is higher inside than outside galaxies. These theories assert that a certain size of fluctuation is needed at 10^{-30} seconds after the Big Bang to begin the galaxy formation process. Inflation, as it turns out, gives fluctuations that are about 100,000 times too large, as well as about 100,000 times too large to be compatible with the measured uniformity of the cosmic microwave background radiation. For these reasons most cosmologists have rejected new inflation and search for yet newer and improved versions.

4. METAFLATION

The inflationary scenario set out to solve two problems, the horizon and flatness problems, and on the road it happened to solve a third, the monopole puzzle. But in reality inflation attempts to do much more. Guth's original scenario showed that, regardless of the initial density of the universe, the present density of the universe should be critical. Any model or theory which shows that the final state of a system is independent of the initial conditions is a very powerful one. If you could prove that inflation took *any* universe—highly chaotic, anisotropic, inhomogeneous—and produced the isotropic universe we observe today, then the scenario would have accomplished a great deal. Because this is the hidden agenda of inflation, many theoretical physicists are highly prejudiced in its favor. Some popular accounts have already implied that inflation is a verified theory. We now ask the awkward question: is the prejudice justified and has inflation in any way been verified? The reader is warned that the following opinions are heterodox and will undoubtedly incur the wrath of our colleagues.

First, we reiterate that the problems inflation set out to solve are of a different sort than predicting the perihelion shift of Mercury or the spectral lines of helium, for which we can experimentally measure the correct result. Rather, the near flatness of the universe strikes us as suspicious, as does the fact that the universe appears isotropic. Unfortunately, there is only one universe at our disposal. We cannot then claim to know with certainty how universes behave in general and hence cannot claim to know that the average universe should be flat or curved, isotropic or anisotropic. We are at liberty to assume the cosmos began flat and isotropic, in which case the questions that inflation addresses disappear. Therefore these issues are philosophical and not physical.

Second, assume that the grand design of inflation is correct: any set of initial conditions before inflation yields the universe we live in. This means, conversely, that by observing the present universe we cannot retrodict and describe the universe before the inflationary period. Any result is valid since all give the present universe. The subject of quantum gravity, for example, which attempts to make statements about the preinflationary epoch must then, by hypothesis, make untestable predictions. Of course, this is not an argument against inflation. It does state, however, that *either* inflation is incorrect *or* some other branch of cosmology becomes metaphysical.

It is appropriate to ask whether inflation fulfills the grand design. While some theorists claim that inflation takes place in anisotropic and inhomogeneous cosmologies, most detailed *calculations* until now have been carried out in isotropic universes. We showed in Section 2 that the horizon problem cannot be solved in an isotropic universe because there is no problem. Thus it is unclear whether these calculations have logically solved anything.

At the time of this writing, researchers have begun to investigate inflation in anisotropic and inhomogeneous cosmologies. The results are to date inconclusive: inflation appears to take place in some models and not in others.

From a complementary point of view, certain of the arguments brought against inflation also appear to be highly untestable. New inflation was thrown out partly because it predicted fluctuations at 10^{-30} seconds that conflicted with a certain model of galaxy formation. There is so much disagreement over the nature of galaxy formation—not to mention the state of the universe at 10^{-30} seconds—that any such argument cannot be taken to rest on ironclad physical principles. Consequently, one unverified model has been rejected because of another unverified model. On the other hand, to discard inflation because it conflicts with observations of the microwave background is a much more concrete reason—*if* you believe that the fluctuations predicted by new inflation could not have been damped out before 100,000 years after the Big Bang, when the microwave background was created.

But is there any experimental evidence for inflation? One can only say no. The correctness of inflation depends primarily on the correctness of the Grand Unified Theories which provide its foundation. The standard GUT model has made the startling prediction that the proton should decay into lighter particles on a time scale of roughly 10^{31} years. Elaborate experiments have been built to measure this phenomenon, but to date proton decay has not been observed. If GUTS should eventually be thrown out, inflation will be thrown out as well.

That the universe should be at exactly the critical density is the one testable hypothesis claimed for inflation and this prediction appears to be incorrect. The metaphysical aspects of the proposals suggested to remedy this situation will become especially evident in what follows.

We said in Section 2 that almost all studies indicate that the current density of the universe is much less than critical. To be more specific, the observed helium abundance can be produced only if the density of baryonic matter (neutrons and protons) is not much more than 10% of the critical value. Studies of the motions of galaxies and galactic clusters might push this number up to 20% or so. One is already faced with a dilemma—either you discard inflation as contradicted by observation, or you postulate some unobserved form of matter that must make up 90% of the mass of the universe. Theoretical cosmologists have chosen the latter path. The most obvious candidates are massive neutrinos which were thought to be discovered in 1980. Even massive neutrinos are not very massive. But they are plentiful: neutrinos outnumber protons and neutrons in our universe by about one billion to one. If these neutrinos had a mass of about 10 electron volts, or 1/50,000th the mass of an electron, the total mass density in neutrinos would be sufficient to close the gap between the observed density and the critical density. However, the 1980 results that claimed a neutrino mass have not been reproduced, and we are forced to conclude that there is little, if any, evidence for such particles.

Theoreticians are not deterred. Grand Unified Theories and their newer counterparts, Susy (*Su*per*s*ymmetric) GUTS, predict that many so-far-undetected particles would have existed at 10^{-35} seconds after the Big Bang. Such entities may have decayed into lighter particles, similar to massive neutrinos, which would inhabit the universe and provide the missing density. We emphasize that there is no evidence for the existence of these particles. You can take the theoretician's point of view: GUT and Susy-GUT particles are soon to be discovered and the entire GUT-inflationary complex will be vindicated. We now show that such a position only makes the problem worse from the astronomer's standpoint.

The Big Bang theory predicts that the universe started at some finite time in the past. The exact age depends on the expansion rate which in turn depends on the density of matter and whether the contents of the universe consist primarily of baryons or radiation (particles that behave essentially like photons and travel at lightlike or near-lightlike velocities). If inflation is correct, and the bulk of the matter in the universe resides in baryons, it is not difficult to show that the age of the universe must be $t = (2/3)H^{-1}$. Here H is the famous Hubble constant which relates a galaxy's distance to the velocity with which that galaxy appears to be receding from us. (See Figure 2.4 for details.)

As students of astronomy know, there is considerable debate concerning the value of H, which must be measured experimentally. H is usually written in units of kilometers/second/megaparsec. Since both kilometers and megaparsecs are distances and cancel each other, this is a rather perverse way of writing 1/seconds, the inverse of a time. Gerard De Vaucouleurs has consistently claimed values for H of over 100 km/s/Mpc, but most investigators believe H is closer to 50 km/s/Mpc. We have seen few claims for H below about 45 and no such estimates have been made in the last decade. To give you a feel for the numbers, $H = 50$ km/s/Mpc translates into about 5×10^{-11} years^{-1}. Its inverse, H^{-1}, is then about 20 billion years and, according to the inflationary scenario, the age of the universe is then $t = (2/3)H^{-1} = 14$ billion years.

Now, surely it is reasonable to assume that the age of stars must be less than the age of the universe. In 1984, A. J. Penny and R. J. Dickens obtained an age of 14–18 billion years for the globular cluster known unimaginatively as NGC 6752 (NGC stands for New General Catalogue). You see that the lower limit of their measurements, 14 billion years, makes the age of this cluster just equal to the age of the universe! This is at best implausible since it must take *some* time after the Big Bang to form stars. Allan Sandage suggests that you need a time of about $(1/5)H^{-1}$. To fit in globular cluster NGC 6752, which is at least 14 billion years old, the age of the universe must be at least $t = 14 + (1/5)H^{-1}$. Using the inflationary formula for t gives $(2/3)H^{-1} \geq 14 + (1/5)H^{-1}$. If you solve this for H, you discover that you can accommodate NGC 6752 only if H is less than about 33 km/s/Mpc, which contradicts all recent measurements.

We have worked this out for a universe in which baryonic matter dominates. But theoreticians are proposing that the as-yet-to-be-detected GUT and Susy-GUT particles in fact dominate the mass density. The particles are relativistic, or behave like radiation. The expansion of the universe proceeds at a faster rate under the influence of relativistic particles than under the influence of baryons, and it is easy to show that an inflationary, "radiation-dominated" universe has an age of only $t = (1/2)H^{-1}$. If $H = 50$,

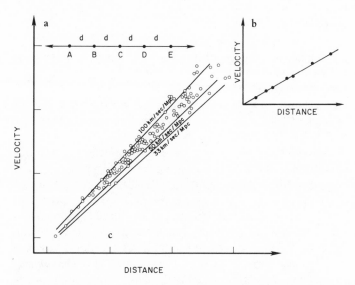

Figure 2.4. The Hubble law. Inset (a) shows a one-dimensional, homogeneous, expanding universe with evenly spaced galaxies A,B,C,D,E. . . . Assume galaxy A observes galaxy B, at distance d, to be receding with a velocity v. By assumption of homogeneity (all points equivalent), B will see galaxy C receding with a velocity also equal to v. Hence, A observes C, at distance $2d$, to be receding with velocity $v + v = 2v$. Any homogeneous, expanding universe should manifest a Hubble law: the recessional velocity of a galaxy should be directly proportional to its distance from the observer, $v = Hd$.

If the real universe were perfectly homogeneous *and* one could accurately measure galactic distances, one could plot v against d for many galaxies to get a graph like (b). The slope of the graph is H, the Hubble constant. The inverse of the Hubble constant, H^{-1}, has dimensions of distance/velocity = time. If you imagine dividing the distance of the farthest galaxy by its recessional velocity, you have essentially divided the size of the universe by its expansion rate. It is not surprising that the resulting time, H^{-1}, is the approximate age of the universe.

The real world is not so simple. A galaxy's recessional velocity is inferred from its redshift, which appears to be a valid assumption. But the distance to a galaxy cannot be measured directly. One indirect approach is to measure the apparent brightness of certain stars in a galaxy. If we assume we know the true brightness of these stars from theory or by measurements of nearby stars, then the distance d can be calculated. Such calculations are fraught with uncertainty and a typical graph of v versus d might look like (c). Which straight line best fits the data? Three possibilities are shown.

this implies that the age of the universe is only 10 billion years and NGC 6752 was formed at least 4 billion years before the Big Bang! To accommodate NGC 6752 in the manner we did above, we must lower H to 21 km/s/Mpc, a value in serious contradiction with all evidence.

At this point, the only way to save inflation—short of throwing out the observations of stellar astronomy—is to modify the equations of relativity. One current suggestion is to reintroduce the infamous cosmological constant into the field equations (see "Garden"). The constant is an arbitrary number added to the equations which can alter the expansion rate and hence the age of the universe. Einstein originally introduced the cosmological constant in order to make his model of the universe static—which he believed the real universe should be. After the universe was discovered to be expanding, he discarded the constant as "the biggest blunder" of his life. Nevertheless, if we assume that the constant exists and has a small value, its effect on the solar system will be undetectable, yet it can lengthen the life of the universe sufficiently to remove the age contradiction in an inflationary universe.

But it turns out that the cosmological constant is effectively the same as the vacuum energy density discussed in the previous section. To assume that today's value of the vacuum energy density is small but nonzero raises further difficult questions—quite apart from the fact that no evidence for it exists except the need to solve the age problem. Specifically, from the viewpoint of GUTS, when the vacuum decays from its metastable state after inflation, there is no known reason why it should drop to exactly zero, a very small nonzero value, or a very large value. If it has a nonzero value, then why is it too small at the present time to be detectable? Notice, this is not more or less of a conundrum than the flatness problem inflation originally set out to solve! If we introduce the cosmological constant to avoid the age contradiction, we merely swap the flatness problem for the vacuum energy density problem: why is the vacuum energy density so close to zero?

Associated with the cosmological constant is a second difficulty: it is a constant. Through Einstein's equations it influences the behavior of spacetime and of every particle in the universe; yet, since it is a *constant*, nothing influences it. Such behavior violates the fundamental concept of action and reaction which lies at the heart of Newtonian mechanics. Einstein incorporated the concept of action and reaction into the theory of relativity by showing that spacetime is not absolute and immutable but is shaped by the objects within it. At the same time, the shape of spacetime determines how objects move. Indeed, Einstein's equations are exactly the equations that show how matter affects the geometry of the universe and vice versa. From this point of view, to introduce a cosmological constant is a very retrogressive step. Action and reaction appears to be a sound principle to demand of a theory: any entity that can affect another entity should itself be affected by this interaction. If the cosmological constant is truly constant, then the principle of action and reaction is violated.

In the above discussion we attempted to show that inflation's solution to the flatness problem brings it into contradiction or near contradiction with actual observations. At the risk of being considered extremists, we might ask whether inflation can *in principle* solve the flatness problem. Suppose inflation takes place and makes the universe extraordinarily—but not exactly—flat. (Inflation does not claim to make the universe *exactly* flat.) The Einstein equations tell us that if the universe is not *exactly* flat then it must always eventually become curved again, though this may take place until the universe is many billions of times older than it is now. So, even if inflation takes place and we wait long enough, the universe will not be flat and the density will not be at the critical density. Therefore, whether the actual density of the universe lies near the critical density depends on the epoch of observation. One might argue that after the universe has aged many billions of times, all stars would have disappeared and civilizations would not be present to make such observations, but this is an anthropic argument (see references to previous essay) and not an inflationary argument. Consequently, inflation only solves the flatness problem if we happen to be living at the proper time.

We leave the reader with the following thoughts. A peculiar situation has arisen in cosmology. Over the last seven years, physicists have been hard at work on a theory which set out to solve two problems that might not exist. The theory has no evidence to support it and the one prediction that it makes appears to be incorrect. In order to reconcile observations with this theory, one must invent new particles that have not been observed. The assumption that these particles exist brings the theory into even more serious conflict with observations unless a further quantity is introduced that also has not been observed. This further quantity brings with it a puzzle that is equivalent to that which the theory was originally invented to solve. Yet because the theory is in some sense pretty, many theorists have been willing to forgo Ockham's razor and accept the added epicycles.

It is of course too early to make a conclusive judgment on inflation that is, without argument, esthetically pleasing. But also without argument is the fact that physics has begun to approach the frontier where it is no longer based on experimental evidence and makes no predictions that are testable. Once this border is crossed we have left the world of physics behind and entered the realm of metaphysics. But at the same time we concede that only twentieth-century scientists recognize the distinction between the two. Perhaps Newton would have agreed with Karl Wojtyla of the Vatican who, having had inflation explained to him, remarked, "That's a good idea, maybe we'll use it next time."

3·GEODESICS, DOMES, AND SPACETIME

1. WORDS

We often use words without giving them much thought. A cynic would go further—it is a rare instance when we *do* think about what we say. There is a case to be made here. If you've ever called anyone a "son of a gun," you probably did not mean he is the offspring of a Colt 45. (The expression seems to have originally denoted "son of a gunner" on a ship. Because sailors were usually out on the high seas, I suspect it meant he was a bastard.) Not long ago, a Scientologist attempted to convince me that the latest Hubbard "Thought-O-Meter," as he called it, worked on the principle of "focusing thought energy from space through the body." "Energy" is a term that by this time has lost the significance it once had and, if you read the next essay in this collection, you might feel forced to agree with Humpty-Dumpty that a word "means anything I want it to mean, neither more nor less."

"Geodesic dome" seems to fall somewhere into this category of used-but-not-understood terms. "Geodesic" signifies "of or pertaining to geodesy," which in turn derives from the Greek *geodaesia, geo* for "earth" and *daesia* meaning "to divide." That is, geodesy relates to the dividing or surveying of the earth. In mathematics, a geodesic has come to mean the shortest distance between any two points on a given surface. I will have more to say about this later. Where do domes fit in? Most of us can picture a geodesic dome in our minds, seen either at world fairs or in countless photographs. But on a moment's reflection it is not entirely obvious that a geodesic dome is any more related to geodesics than a dome of the more pedestrian sort. Thus it is prudent to reflect a little longer.

2. HISTORY

Most books on unconventional housing construction are not written for the scientist and are consequently not very instructive when it comes to the mathematics behind geodesics and domes. But the history of the structure is quite unexpected and turns out to be intimately connected with the development of the planetarium.[1]

Eudoxus of Cnidos (c. 390–c. 340 B.C.) was, next to Archimedes, the greatest mathematician of ancient Greece. He invented the *method of exhaustion*—today called the *limiting process*—which will be relevant to what I have to say later. He also authored the first known theory to explain in geometric terms the motion of the planets and is thus considered the founder of modern astronomy. His model of the solar system placed the planets in concentric spheres, the combined motions of which were meant to explain the observed movements of the planets. It is sometimes said that he also built the first model of the heavens, a celestial globe like those found in today's classrooms, and took it out of Egypt around 370 B.C. I have not found any convincing evidence for this and wonder if there is a confusion between the term "model" as in a scientific theory, and "model" as in a physical construct such as a celestial globe.

Nevertheless, his theory of the heavens did inspire the Greek poet Aratus to write a very long and tedious poem called *Phenomena*, which is a versified description of the universe according to Eudoxus. Aratus starts with an invocation to Zeus:

> *From Zeus let us begin; him do we mortals never leave unnamed.*

He continues for fifty pages running through the constellations:

> *Andromeda, though she cowers a good way off, is pressed by the mighty Monster of the sea.**

* Cetus.

He eventually gets around to weather prediction:

> *Mice in the daytime toss straw and are fain to build a*
> *nest when Zeus shows sign of rain.*

And he ends with an exhortation to

> *Study all the signs together throughout the year and*
> *never shall thy forecast of the weather be a*
> *random guess.*[2]

Meteorologists take heed. *Phenomena* had much in common with today's bestsellers: it was badly written and phenomenally successful. Cicero translated it into Latin, and to illustrate the poem media tie-ins in the form of celestial globes were manufactured and sold. The globes have since been called Aratus globes and are the first known models of the heavens definitely known to have been constructed. Consequently, one might argue that a poet was responsible for the original planetarium.

Over the centuries globes and orreries became larger and more elaborate, but all had one thing in common: the viewer observed the heavens from the outside. It is unnatural, if not entirely preposterous, to view the heavens from outside the universe, yet this was the situation. Matters finally changed in 1664 with the Gottorp Globe designed by Adam Olearius in Germany. Here was a water-powered, 3½-ton, 10-foot sphere which rotated every twenty-four hours to imitate the real sky. The most remarkable thing is that up to twelve people could sit on a platform built *inside* the globe and gaze at the gilded stars above their heads. The Gottorp Globe (Figure 3.1) is now housed in the Lomonosov Museum in Leningrad. A similar device was constructed in 1758 in Cambridge, England, by Roger Long; it consisted of a globe 18 feet in diameter and accommodated thirty people. Long's globe had small holes of various sizes punched in it in the proper locations to represent the stars and constellations. It never became very popular and was sold for scrap in 1874. Several other globes have been constructed along these lines. The most recent was designed in 1911 by Dr. Wallace Atwood of the Chicago Science Museum. This globe (see Figure 3.2) is 15 feet in diameter, electrically driven, and still in use today.

Figure 3.1. The Gottorp Globe was designed by Adam Olearius, the court mathematician to Duke Frederick of Holstein-Gottorp. It was built at the duke's castle between 1654 and 1664 by Andreas Busch and is 3.1 meters (10 feet) in diameter. The globe was given to Tsar Peter the Great of Russia in 1715. It was badly damaged by fire in 1747, rebuilt and modernized in the following five years, and is currently on display at the Lomonosov Museum in Leningrad. The stars and constellations were painted on the inner surface. The Gottorp Globe could accommodate ten spectators who crawled in through a hole in the Indian Ocean and sat on a platform inside. (Courtesy of Carl Zeiss, Inc., Thornwood, N.Y.)

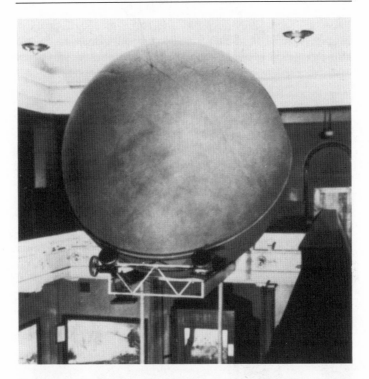

Figure 3.2. The Atwood Celestial Globe is the largest hollow star globe still in operation. The globe was designed by Wallace A. Atwood, director of the Chicago Science Museum. It is 4.57 meters (15 feet) in diameter and is illuminated from the exterior. The light shines through 692 holes punched in the surface which depict stars down to the fourth magnitude and some of the fifth. (Courtesy of Carl Zeiss, Inc., Thornwood, N.Y.)

One cannot keep on building bigger and bigger spheres in this fashion to accommodate bigger and bigger audiences—rotating the building becomes too difficult and is an inefficient use of space. This was the problem faced in 1913 by the Carl Zeiss Optical Company of Germany, which wanted to build a large planetarium capable of accommodating equally large audiences. The solution was not found until after World War I, when Dr. Walter

Bauersfeld, then chief engineer of the firm, made a Copernican suggestion: hold the building stationary and fashion a movable projector which would cast the images of the stars and constellations onto the inner surface of the dome. Thus was born the modern planetarium.

Bauersfeld was primarily concerned with the optics and the geometry of the projection system (Figure 3.3). Yet the Zeiss Company required a dome—one that would not only be large but lightweight as well, because the plans called for it to be placed on the roof of the Zeiss factory in Jena, now locted in East Germany. After several years, Bauersfeld and his team produced the final design: a hemispheric dome, 16 meters in diameter, braced by a light iron framework consisting of 3,840 struts, each 60 centimeters in length and accurate to .06 millimeter (Figures 3.4–3.6). The hemispheric shape was derived from a truncated icosahedron (see Section 4 to learn how this works) and covered with a thin shell of concrete.[3] The thickness of this shell was determined by using the ratio of the thickness of an eggshell to an egg. In August 1923 the Zeiss planetarium and also the world's first geodesic

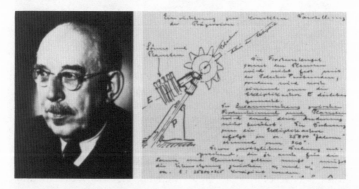

Figure 3.3. Walter Bauersfeld (1879–1959). Former director of the Carl Zeiss Optical Company, inventor of the planetarium projector and the geodesic dome. He is pictured here with a page of his manuscript showing the first planetarium projector. (Courtesy of Carl Zeiss, Inc., Thornwood, N.Y.)

Figures 3.4, 3.5, 3.6. The world's first geodesic dome under construction atop the Zeiss factory in Jena, Germany, 1922. The 16-meter dome was designed by Bauersfeld and constructed by the firm of Dyckerhoff and Widmann. Figure 3.5 shows the dome after it had been covered with a thin concrete shell. Figure 3.6 presumably shows the grand opening, August 1923. (Figures 3.4 and 3.5 courtesy of Dyckerhoff and Widmann archives, Munich. Figure 3.6 courtesy of Zeiss, Inc.)

dome were opened to the public. The planetarium became known as the "Wonder of Jena."

The Zeiss Company built other planetariums but apparently none was housed in domes we would term geodesic. The firm of Dyckerhoff and Widmann, which actually covered the skeleton with its ferroconcrete shell, continued to perfect the technique on ever larger frameworks, but here too none seems to have been "geodesic." Judging from an article by Dischinger,[4] the main concern of the time was not the strength of the dome but the ease with which it could be assembled and sprayed with concrete. Dyckerhoff and Widmann found other configurations more desirable and by 1941 had roofed over 1,600,000 square yards by shell construction. In 1938 both the Zeiss Company and Dyckerhoff & Widmann were awarded the Franklin Institute Medal for the shell-form construction of planetarium domes.

Was the Zeiss dome patented? This is an interesting question. A letter from W. Degenhard of Zeiss to Shelter Publications would indicate no. From this letter I also infer that the Wonder of Jena was destroyed during World War II. Degenhard writes:

> It is very difficult to find the proper answers to your questions. You have to consider that Carl Zeiss in West Germany had to be built from scratch. We have no access to the archives in East Germany. All patents and recordings were taken away either by the American or the Russian armies when they occupied Jena in 1945. Dr. Bauersfeld was among the 126 people who were brought to West Germany by American troops in 1945.
>
> . . . We were not able to find any patent which covered the subject of the planetarium or the dome. At that time it was the principle of our firm, that such basic findings should be made available to the whole scientific world, so it may well be that nothing was patented in Germany.[5]

The sentiments expressed by Degenhard are genuinely touching and I wish the real world were as he portrays it. But I believe the Zeiss Company was somewhat more practically minded and did patent its dome.

The first clue is an oblique sentence in Dischinger's article that reads, "The further development of this patented dome construction then followed the collaboration of the two companies." Dischinger does not tell us who received the patent or for what—for the dome or for the technique of spraying ferroconcrete. Nonetheless, the sentence is sufficient to arouse one's curiosity, as it did mine. I recently wrote to Dyckerhoff and Widmann, who had sent me Dischinger's article, asking for further information. They kindly obliged by sending me a copy of the patent itself (see Figure 3.7). The document is dated 1925 and the patent claim is for a "method for the fabrication of domes and other curved surfaces of reinforced concrete." The language of the patent is not very precise but is clearly made out to the Zeiss Company of Jena and clearly refers to the technique used to build the Jena dome. To a logician it follows that the patent is for a geodesic dome. But law is not logic and I am not a lawyer; therefore I will comment no further on the matter. The term "geodesic" was applied to domes by Buckminster Fuller, who received a U.S. Patent in 1954.

3. GEODESICS

A geodesic is the shortest distance between two points. On a flat, two-dimensional surface, such as a sheet of paper, a geodesic will simply be the straight line joining these two surfaces. The same holds true in three dimensions as well if the space is Euclidean (Figure 3.8a,b,c). By Euclidean space I mean any space or surface of any dimension where Euclidean geometry is known to hold. Euclidean geometry is the geometry we all learn in high school and is often termed plane geometry, but I will avoid using that term. In Euclidean spaces the interior angles of triangles always add up to 180°, and the Pythagorean theorem can be written in the familiar form $s^2 = a^2 + b^2$, where s is the length of the hypotenuse and a and b are the lengths of the other two sides. Using this formula one can easily calculate the length of any geodesic in Euclidean space, as is also shown in Figure 3.8. I do not want to call Euclidean geometry "plane geometry" because, although all Euclidean spaces are flat, not all flat spaces are Euclidean. This will

DEUTSCHES REICH

AUSGEGEBEN
AM 19. JUNI 1925

REICHSPATENTAMT

PATENTSCHRIFT

№ 415395

KLASSE 37a GRUPPE 2

(Z 13458 V/37a)

Firma Carl Zeiss in Jena.

Verfahren zur Herstellung von Kuppeln und ähnlichen gekrümmten Flächen aus Eisenbeton.

Patentiert im Deutschen Reiche vom 9. November 1922 ab.

Die vorliegende Erfindung betrifft ein Verfahren zur Herstellung von Kuppeln und ähnlichen gekrümmten Flächen aus Eisenbeton, das sich durch besondere Wohlfeilheit auszeichnet. Das neue Verfahren besteht darin, daß ein sich selbst und einen Teil des Gesamteigengewichts tragendes, in der Dachhaut liegendes räumliches Netzwerk aus Eisenstäben aufgebaut und unter Verwendung leichter, unmittelbar an das Eisenwerk angehängter Schalungen, beispielsweise durch das Spritzverfahren, mit dem zur Erreichung der vollen Tragfähigkeit erforderlichen Betonmantel umhüllt wird. Man braucht nur einen kleinen Teil der Schalung auszuführen und nacheinander an alle Stellen der gekrümmten Fläche zu bringen. Dabei ist diese Teilschalung so an dem Eisengerippe zu verteilen, daß die Eisenstäbe keine wesentlichen Biegungsspannungen bei der Herstellung des Betonmantels erfahren. Durch die Verwendung des bekannten Spritzverfahrens werden neben der Erhöhung der Betonfestigkeit die sonst unvermeidlichen Erschütterungen und Belastungen des Traggerippes und der Schalung bei der Herstellung des Betonmantels vermieden. Bei der Anwendung des neuen Verfahrens erhält daher das Netzwerk auch während der Ausführung des Baues keine nennenswerten Biegungsbeanspruchungen; infolgedessen ist auch bei großen Spannweiten der Aufwand von Eisen verhältnismäßig sehr gering. Ferner wird bei dem neuen Verfahren eine kostspielige Unterrüstung vermieden, an deren Stelle die erwähnten Schalungen treten, und es fallen auch die Ausrüstungsspannungen so gut wie vollständig weg, denen sonst bei der Bemessung der Stärke der Einzelteile Rechnung getragen werden muß.

PATENT-ANSPRUCH:

Verfahren zur Herstellung von Kuppeln und ähnlichen gekrümmten Flächen aus Eisenbeton, dadurch gekennzeichnet, daß ein sich selbst und einen Teil des Gesamteigengewichts tragendes, in der Dachhaut liegendes Netzwerk aus Eisenstäben aufgebaut und unter Verwendung leichter, unmittelbar an das Eisenwerk angehängter und entsprechend dem Fortgang der Arbeiten zu verschiebender Schalungen durch das Spritzverfahren mit dem zur Erreichung der vollen Tragfähigkeit erforderlichen Betonmantel umhüllt wird.

Figure 3.7. The patent for the Zeiss dome. The patent claim at the end of the document reads in translation, ''Method for the fabrication of domes and other curved surfaces of reinforced concrete. The method is based on a spatial network of iron bars which bears its own weight as well as part of the total weight of the concrete. A lightweight form is placed behind the network while spraying the network with concrete, thereby implanting the network in concrete and giving the shell its full strength.'' The bulk of the patent elaborates on this slightly, but provides no detail on the nature of the ''network.''

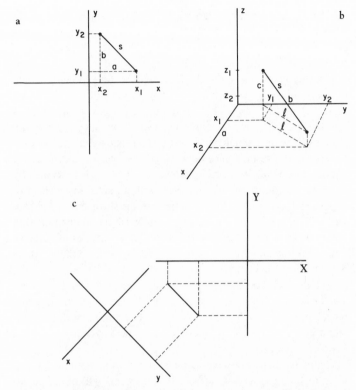

Figure 3.8. The Pythagorean theorem and distance. (a) Say a house is located at coordinates (x_1,y_2) and a mailbox at (x_2,y_2). The difference in the x-coordinates between the house and mailbox is $x_2 - x_1 = a$. Similarly, the difference in the y-coordinates is $y_2 - y_1 = b$. The Pythagorean theorem tells us that $s^2 = a^2 + b^2$. The distance between the house and the mailbox, s, is then $s = a^2 + b^2$. (b) The same holds true in three dimensions. By the Pythagorean theorem for two dimensions, $l^2 = a^2 + b^2$ and $s^2 = l^2 + c^2$, where c is the difference in the z-coordinates. This gives $s^2 = a^2 + b^2 + c^2$. Analogous expressions hold in any number of dimensions. (c) The distance between the house and mailbox does not depend on the coordinate system X,Y or x,y. Whether you choose the origin of the coordinates at the supermarket or in another town, no matter which direction the axes are pointed, you will always measure the same distance s between the house and the mailbox. Any quantity which is independent of coordinate systems is termed an *invariant*.

become apparent in Section 5 when I discuss special relativity. In modern physics and mathematics one often hears the term "Euclidean manifold," which is just an aristocratic designation for a flat space where the geometry of Euclid is valid.

But there is no reason to think that Euclidean geometry should be the only valid geometry. On a sphere the rules of Euclidean geometry do not apply. The first thing one notices in Figure 3.9 is that the triangle contains more than 180°. And since there are *no* straight lines on a sphere, it is pretty clear that a straight line cannot be the shortest distance between two points. Here, the role of geodesics is taken by "great circles." A great circle is formed by slicing the sphere through its center with a plane, say a sheet of paper. The intersection of the plane with the surface of the sphere will be a great circle. The earth's equator is a good example of a great circle, as are all the circles of longitude. Circles of latitude, other than the equator, are *not* great circles because their centers do not lie at the center of the earth. Everyone has heard a pilot announce, "We are flying a great circle route between New York and London." He means exactly what he says: the shortest distance between New York and London goes north and then south, and it is not the first line one might be tempted to draw to connect them on a flat map.

Geometry does not have to be confined to spheres or circles. Each different space (or surface, if you prefer) will have its own rules. The geodesics will be neither straight lines nor circles but figures which have no common name and are described by complex equations. The angles of a triangle may sum to more than 180° or may sum to less, depending on the shape of space. However, such details are not needed for the sake of this discussion. "The geometrical mind is not so closely bound to geometry that it cannot be drawn aside and transferred to other departments of knowledge," said de Fontelle. We will follow his advice and transfer to more practical pursuits.

4. DOMES

A geodesic dome may be constructed in a number of ways. I will not concern myself with the more exotic strategies that proceed

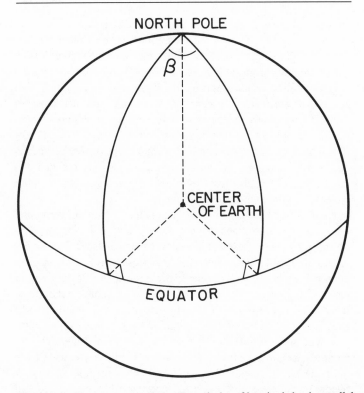

NORTH POLE

CENTER
OF EARTH

EQUATOR

Figure 3.9. Geometry on a sphere. Two circles of longitude begin parallel to each other at the equator but intersect at the north pole. Thus, on a sphere, parallel lines may intersect. Also, because the circles of longitude intersect the equator at 90°, the triangle drawn contains interior angles summing to 180° plus the third angle β. This means that unless the circles coincide and β is zero, a triangle on a sphere will always contain more than 180°, in distinction from a Euclidean triangle which contains exactly 180°.

from such chimeras as "rhombic tricontahedrons." Nonetheless, a geodesic dome is a geometric object, and the architect who plans to build one must start with geometry. The method used by Bauersfeld and still often practiced today is based on the "truncated icosahedron." An icosahedron is a twenty-sided polygon, as shown in Figure 3.10a. Each side or face is triangular and the twelve vertices of the icosahedron are formed each time five of the triangular faces come together. Thus, each vertex has five lines or edges radiating from it. If we slice off the vertices by cutting the edges in thirds (Figure 3.10b) the result will be an object that somewhat resembles a soccer ball. This is the truncated icosahedron. Where the five triangles meet to form a vertex is now a pentagon, and a hexagon now replaces each triangular face. Therefore, the truncated icosahedron is composed of twelve pentagons and twenty hexagons. Because the vertices have been sliced off, the truncated icosahedron obviously resembles a sphere more closely than the untruncated icosahedron.

Now, if lines are drawn from the center of each pentagon and hexagon to their vertices (Figure 3.10c), we will end up with an object that is once more composed of triangular faces, but the triangles are smaller than on the original icosahedron. This last operation does not make the object any rounder but, if we think in terms of engineering, it will make the structure more rigid because the triangle is one of the most rigid shapes known. Finally, if you start from the center of any hexagon now subdivided into triangles, you will be able to travel all the way around the "globe" without changing course. This will not be true if you start from the center of a pentagon. The fact that many of the lines now completely encircle the object is also important from an architectural point of view. Such "great circles" (here made up of straight line segments) distribute stresses throughout the structure rather than in one isolated area. If the subdivided-truncated icosahedron is now cut in half, the result is a geodesic dome. Or two.

One does not have to stop here. The vertices of the subdivided-truncated icosahedron can again be lopped off, producing more pentagons and hexagons, then subdivided again with shorter struts. The process can be repeated infinitely and the structure becomes rounder each time. Now the reader can guess why the half-subdivided-truncated icosahedron is called a geodesic dome. Mainly, I think, because no one in his right mind would call it a half-subdivided-truncated icosahedron. The other rationale is that the straight line segments which encircle the dome *approximate*

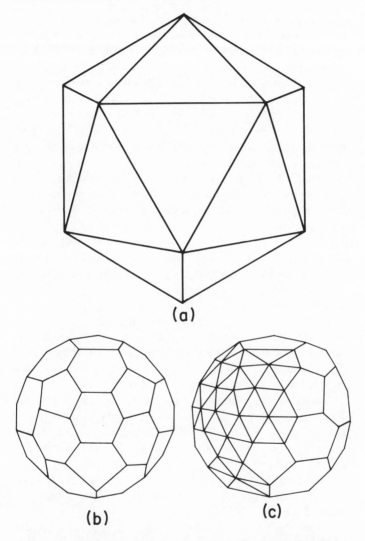

Figure 3.10. Steps to a geodesic dome. Start with a regular icosahedron (a) which has twenty triangular sides and twelve vertices. Slice off the vertices to get a soccer-ball object with 12 pentagonal faces and 20 hexagonal faces (b). Divide the pentagonal and hexagonal faces into triangles to get a "geodesic sphere" (c). Cut in half to get two domes. Serve fresh.

great circles or geodesics. It also turns out, not too surprisingly, that if you put a light bulb in the center of the dome and project the shadows of the struts on the inside of a real sphere, the shadows will be exactly geodesics.

So we see that the term ''geodesic'' dome is correct only if it merely signifies ''earth dividing,'' for that is what has been done. However, ''geodesic'' here refers to the shortest distance between two points. In this case it must be less correct than calling an ordinary hemispherical dome a geodesic dome, because on the latter great circles can truly be drawn, whereas the ''great circles'' on ''geodesic'' domes are actually composed of straight lines.

But let us not be too dogmatic. ''Approximately correct'' is often good enough in mathematics and physics if for no other reason than that's all you can hope for. In difficult problems exact solutions are the exception and approximations the reality. How closely does a geodesic dome approximate a truly hemispherical dome? The answer is, as closely as you want. To demonstrate this, let me consider a related question that is at least three thousand years old: what is the value of π?

Today we know that π is an irrational number so its true value cannot be calculated exactly. But it can be approximated. The classic method was devised by Archimedes (c. 297–212 B.C.) and is illustrated in Figure 3.11 and the accompanying box. We see a circle bracketed by two hexagons, one smaller than the circle and one larger. Using elementary trigonometry—or if you're Archimedes, the Pythagorean theorem—it is easy to show that π must lie between 3.0 and 3.46. If the circle is not similarly bracketed by two dodecogons (twelve-sided figures), it is no more difficult to show that π must lie between 3.106 and 3.215. If we choose a ninety-six-sided figure, the same process yields a value for π between 3.141 and 3.143, which is now very close to the accepted value of 3.141592. . . .

The relevancy to the question of geodesic domes is clear. As the number of sides of the polygon increases, the length of each side becomes shorter and the perimeter approaches that of a circle. When does a polygon become a circle? When the number of sides becomes infinite and the length of each side shrinks to zero. When does a geodesic dome become a hemisphere? When the number of struts becomes infinite and the length of each strut goes to zero. In other words, when you can calculate π exactly. That is, never. But you can get as close as you want. Here is the notion of a limit, invented by Eudoxus, who also gave us the first celestial model.

5. SPACETIME

I have sketched the evolution of the planetarium and showed how it interesected paths with geodesics at the point where a practical need for a large dome arose. Geodesics are found not only in architecture but in nature as well. Newton's first law of motion states that any object not acted upon by an outside force will travel in a straight line. Newton considered space to be Euclidean, so his first law is merely a statement that all objects tend to follow geodesics unless something prevents them from doing so. In optics, Snell's law relates the angle of incidence of a light beam into a crystal to the angle of refraction. The resulting mathematical expression is the statement that, while entering and exiting the crystal, light travels along the shortest possible path—a geodesic.

There are doubtlessly many other example of geodesics in nature which I haven't thought about. But one subject where the notion of geodesics finds a natural and central role is the theory of relativity. I have already discussed the Pythagorean theorem in two dimensions, where $s^2 = a^2 + b^2$, and Figure 3.8b showed how this could be extended to three dimensions, where $s^2 = a^2 + b^2 + c^2$. There is no reason to stop there, and we could equally well write $s^2 = a^2 + b^2 + c^2 + d^2$ in four dimensions or $s^2 = a^2 + b^2 + c^2 + d^2 + e^2$ in five. The human imagination has difficulty with four dimensions and higher, but the beauty of mathematics is that it can extend the imagination beyond the realm of everyday experience.

The formulas I have written down are valid only when space—being two, three, or four dimensions—is flat and Euclidean. In a Euclidean space Newton's first law is valid. Unimpeded objects will travel along straight lines, and thus two rockets launched on parallel courses will remain parallel and never collide. We say

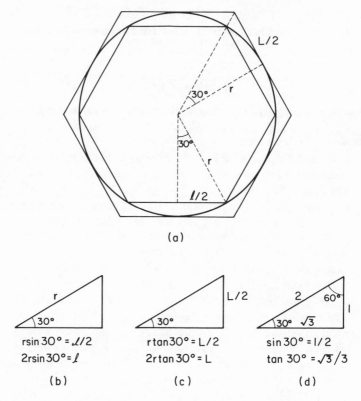

(a)

$r \sin 30° = l/2$
$2r \sin 30° = l$

(b)

$r \tan 30° = L/2$
$2r \tan 30° = L$

(c)

$\sin 30° = 1/2$
$\tan 30° = \sqrt{3}/3$

(d)

Figure 3.11. Archimedes' method for calculating π. See accompanying box.

ARCHIMEDES' METHOD FOR CALCULATING π

The circumference of the circle C should be greater than the perimeter of the inner hexagon P_i and less than the perimeter of the outer hexagon P_o. In symbols,

$$P_i < C < P_o.$$

Now, from Figure 3.11a, $P_i = 6\ell$, where ℓ = length of side of inner hexagon. Also from Figure 3.11b we have

$$P_i = 6\ell = 12r\sin 30°.$$

Similarly, from Figure 3.11c we have

$$P_o = 6L = 12r\tan 30°.$$

Since $C = 2\pi r$, this implies

$$12r\sin 30° < 2\pi r < 12r\tan 30°.$$

Dividing by 2r gives

$$6\sin 30° < \pi < 6\tan 30°.$$

Since we are dealing with 30-60-90 triangles, Figure 3.11d allows us to evaluate this to get

$$6 \cdot (1/2) < \pi < 6 \cdot \sqrt{3}/3.$$

Or

$$3 < \pi < 3.46.$$

If the polygon has n sides and $\theta = 360/n$, then

$$n\sin \theta/2 < \pi < n\tan \theta/2.$$

their world lines never intersect. Euclidean spaces have their uses in modern physics but I will not be concerned with them in the remainder of this essay. I will be concerned with one type of four-dimensional manifold—spacetime. Spacetime has three spatial dimensions and one time dimension and is the arena of the theory of relativity.

The reader probably knows that there is a special theory of relativity and a general theory of relativity. The special theory assumes that planets, stars, and galaxies do not exist and that spacetime is flat. Moving spaceships follow straight lines. General relativity is more accurately termed a theory of gravitation. It allows stars and galaxies to enter the picture and asks how their gravitational field affects spaceships, other stars and galaxies, and the universe itself.

For instance, suppose two spaceships were launched on nearby parallel courses in otherwise empty space. This is the domain of special relativity, and we expect the ships' paths to be straight lines and nonintersecting. Now suppose, however, that the ships are very large and massive, say like an Imperial Battle Cruiser from *Star Wars*. Eventually the mutual gravitational attraction of the ships would pull them together until they collide. *Their two world lines began parallel but eventually intersected.* We know from Euclidean geometry that two parallel lines never intersect, so what happened here? If you glance again at Figure 3.9, you will notice that on the surface of a sphere, two initially parallel lines can intersect—for instance, as the lines of longitude intersect at the north and south poles.

This analogy is exactly how the term "curvature of space" found its way into relativity. You can say the gravitational attraction of the two spaceships pulled them together, or you can say that their gravitational fields curved space and that the ships no longer followed straight world lines but curved geodesics. When the spaceships became large enough so that their gravitational fields were no longer negligible and their world lines no longer straight, we left special relativity and entered the province of general relativity. The amount at any time by which the ships' world lines tilted toward each other is called "geodesic deviation" or, in more common language, tides. The distance above the average height of the ocean that the moon raises tides is a measure of how much space is curved in our vicinity.

It is fair at this point to stop and question whether this geometric terminology is really necessary to interpret relativity. The following example might suggest otherwise.

Recall that $s = \sqrt{a^2 + b^2 + c^2}$, found by the Pythagorean theorem, is the distance between any two objects in Euclidean space. The distance s is independent of any reference frame or coordinate system in which we choose to measure it (Figure 3.8c) and is thus said to be *invariant*. In real life, however, events are separated in time as well as in space. Two firecrackers exploding at midnight and at 12:01 A.M. at opposite ends of the street have not only different spatial coordinates but different time coordinates as well. You may have heard that in relativity, distances and times contract and dilate to moving observers. This is true. Both the space separation and the time separation between events depend on the observer's frame of reference—for instance, on how fast he is moving—and ordinary distances and times are no longer invariant. What turns out to be invariant is the quantity $s^2 = -t^2 + a^2 + b^2 + c^2$, which is equivalent to the Pythagorean theorem for special relativity and is often called the Lorentz or Minkowski metric in honor of the two physicists who extensively explored its properties early in this century. Here is an example of a flat space that is not Euclidean.

Sometimes $s = -t^2 + a^2 + b^2 + c^2$ is termed the spacetime distance between two events. But we see that if $t^2 = a^2 + b^2 + c^2$, then $s = 0$, and if t^2 is greater than $a^2 + b^2 + c^2$, then s^2 is a negative number and s itself is imaginary. To speak of imaginary distances is not very helpful. It is this sort of difficulty with spacetime geometry which has led certain physicists, notably A. P. French and Nobel laureate Steven Weinberg, to abandon the geometrical approach to general relativity. They argue it doesn't make much sense to talk about time as a fourth dimension; neither does it make sense to speak of imaginary distances or curvature or geodesics and that the geometrical interpretation of relativity has "dwindled to a mere analogy." What is important, Weinberg writes, "is to be able to make predictions about images on the astronomers' photographic plates, frequencies of spectral lines and so on, and it simply doesn't matter whether we ascribe these predictions to the effects of gravitational fields on the motion of planets and photons or to a curvature of space and time. However, Weinberg concedes, "The reader should be warned that these views are heterodox and would meet objections from many general relativists."[6]

Weinberg has a point. I did indeed introduce the curvature of space above as an analogy. But if geometry is a "mere analogy" it is a very useful "mere analogy." The equations or relativity which describe the motion of objects are almost identical to those of classical differential geometry, and the geometric way of thinking often provides shortcuts in calculations which save much time in comparison to the methods of which Weinberg is so fond.

If you are not convinced, perhaps a further example will help make the geometric case. The gravitational field of a black hole, for instance, may curve space greatly, in the sense that geodesics will deviate considerably from straight lines. If a relativist wanted to calculate what happened to the fabric of spacetime when two black holes collided, he would have a difficult job because the equations describing the shape of space in this situation are horrendously complicated. Actually, the equations of relativity are so horrible that none but the simplest problems can be solved exactly. The answer is generally obtained by approximation on a computer.

So then, how do we approximate the shape of spacetime? One way is to guild a geodesic dome. This is the philosophy behind the branch of relativity known as Regge calculus, first developed by Tullio Regge in the late fifties and early sixties. Earlier I talked about how an icosahedron could be used to approximate the surface of a sphere and that, if the icosahedron was truncated, the approximation became better. The surface of a sphere is a two-dimensional manifold which lies in our ordinary three-dimensional space. A geodesic dome is also a two-dimensional surface which is "embedded" in three-dimensional space. The difference between the sphere and the dome is that the sphere's surface is curved but that the dome is composed of Euclidean triangles. The approximation to the curvature comes at the joints between the triangles.

Regge calculus extends this idea to spacetime. Spacetime may be viewed as our three-dimensional space moving through time. In other words, a three-dimensional manifold embedded in a four-dimensional space. To approximate the shape of spacetime, we build three-dimensional geodesic domes (or similar structures) and evolve them in time. A three-dimensional geodesic dome will be much like the ordinary type except that the struts will no longer lie only on a two-dimensional surface but also form a framework inside. In 1970 Cheuk-Yin Wong used Regge calculus to model a spherically symmetric, electrically charged black hole. The curvature of space was approximated using icosahedrons. If the icosahedrons had been truncated, greater accuracy could have been achieved. What kind of accuracy? How accurately can you approximate π?

Regge calculus is a clear link between geometry and relativity. Unfortunately, it is rather difficult to implement and not many large-scale calculations have utilized it. Yet very similar procedures are used, for instance, in the study of quantum gravity. Einstein's original theory of gravitation is referred to as "classical," meaning that no quantum mechanical effects are taken into account. Quantum mechanics holds that when one looks at small enough scales many phenomena are not truly continuous as they appear in the large but only take place in discrete intervals. A bicycle wheel, for example, can rotate only at discrete frequencies, but the frequencies are so closely spaced that the motion appears continuous. In atoms the situation changes. Electrons can orbit the nucleus only at discrete frequencies which are easily measured in the laboratory. So shortly after the Big Bang, before the universe was 10^{-43} seconds old and when measurable distances were only 10^{-33} centimeters, quantum mechanics should have been very important. If present ideas are correct, it might be that spacetime itself was quantized, that is, not continuous as it appears to us, but discontinuous on a scale of 10^{-33} centimeters. Sizes smaller than this so-called Planck length did not exist. Spacetime appeared not like a smooth sphere or saddle but like a window screen. Or a geodesic dome. For this reason some investigators are modeling quantum theories of gravity on computers with geodesic-domelike structures in which the struts are 10^{-33} centimeters long. Don't ask me what is between the struts; nobody knows.

Regardless of whether quantum gravity turns out to have any bearing on reality, almost all computational methods in relativity have their roots in geometry. A typical procedure is to specify the geometry on one slice of three-dimensional space, then tell the computer how to evolve through time. In such procedures, quantities such as "the extrinsic curvature" and "three-scalar curvature" of the manifold are essential.

To conclude this essay, I would maintain that if geometry is merely an analogy for gravitation, it is an analogy without which relativists would never get any work done. More importantly, I hope the reader sees how the evolution of a few ideas—geodesics and limits—links such diverse fields as architecture and cosmology. This is the way science works: the simplest ideas have the widest application. And it is entirely appropriate that the first architectural application of geodesics should be in a structure so intimately connected with cosmology where the concept of a geodesic is essential. I do not think it is coincidental. Neither am I surprised that Bauersfeld, a man trained in engineering and astronomy, thinking like the ancient Greeks about ways to divide the earth, became the inventor of the geodesic dome. There is also a hidden moral lurking in this history; it is a moral which tells us not to worry about being too practical. Who would have thought a poem could lead to the planetarium?

4·THE EVOLUTION OF ENTROPY

*The metaphor is probably the most
fertile power expressed by man.*

—Jose Ortega y Gasset, 1948

1. THE ENTROPY OF METAPHOR

"Metaphor is the frayed thread that connects what we desire with what merely exists," scribbled Tony Rothman in 1987. Neither Gasset's metaphor nor my own is particularly powerful. Equating *metaphor* with *fertile power* does not require a great leap of the imagination. My draft needs to be tightened up; it holds promise but is too vague to be really effective and vivid.

Nevertheless, many of you would probably agree with the above sentiments. A good metaphor discloses a link between two or more concepts that we had previously thought unrelated and, in doing so, broadens our experience of the world. The better the metaphor, the deeper and more striking the connection.

Surely the grand master of verbal allegory was Shakespeare, who scattered merely good metaphors as a benevolent monarch scatters goldpieces. Typical:

> *Your face, my thane, is as a book where men
> May read strange matters*
> *(Macbeth, 1.5.63)*

which, though typical, resonates much more deeply than "You look disturbed."

The modern reader may not register

> *Thy head is as full of quarrels
> as an egg is as full of meat*
> *(Romeo and Juliet, 3.1.23)*

75

but not infrequently we encounter a really superb figure of speech:

Glory is like a circle in the water,
[Here you are puzzled; what possible connection could
    exist between glory and a circle in the water?]
Which never ceaseth to enlarge itself,
[Already very good; both glory and water rings expand.]
Till by broad spreading it disperse to naught.
[Now this is superb. He's managed to connect *two*
    complementary properties of water rings to glory.
Who could have foreseen it?]

(*Henry VI*, Part I, 1.2.133)

The metaphorical content of language has diminished since Shakespeare's time but man's desire to find the metaphor for existence has not. And since science and technology have become dominant forces in contemporary society, it is not surprising that writers, artists, and scientists themselves have turned to science for the metaphor of metaphors.

The most popular choice must be Einstein's theory of relativity, which has influenced everything from twentieth-century art to thinking on cultural ethnocentricity. Not far behind is Heisenberg's Uncertainty Principle, a sort of quantum-mechanical Murphy's Law which explains, if not why everything goes wrong, then at least why you can't get everything right. And not to be outshined is the second law of thermodynamics, which explains in accordance with Hindu mythology why things fall apart.

Things fall apart. The End is Nigh. Order versus chaos. The second law has become our metaphor for complexity beyond comprehension, decay, chaos. In other words, modern life. Is the metaphor valid? Or is it merely a catchword like *kharma*, which has descended from the Indian *deed, work, action* to the point where "good kharma" means about as much as "Kharma-Cola"? That is the subject of this essay. Ironically, while a philosopher, Gasset, expressed a poet's faith in the power of metaphor, it was a poet, Paul Valéry, who sounded a scientist's warning against the misuse of metaphor: "The folly of mistaking a paradox for a discovery, a metaphor for a proof, a torrent of verbiage for a spring of capital truths, and oneself for an oracle, is inborn in us." Take heed.

2. THE ENTROPY OF THERMODYNAMICS

Unlike psychologists, physicists are not very metaphorical and their jargon is disappointingly uninspired and penetrable. The term "thermodynamics" means exactly what is says: it is that branch of physics which deals with the dynamics of heat. Dynamics, from the Greek word for "power" or "strength," connotes power to set things in motion. So thermodynamics concerns the motion of heat. You might well wonder how something as abstract as heat can be in motion, but historically this was a very natural way of looking at things. To early nineteenth-century scientists, heat was a fluid, often called caloric, that could flow from a hot body (usually termed a heat source or reservoir) to a cold body (termed a heat sink) and in the process be made to do useful work.

The most prevalent example was the steam engine, which had appeared late in the eighteenth century. A heat source (burning coal) produces steam in a boiler which is then injected into a cylinder housing a piston. The pressure of the steam drives the piston to the top of the cylinder. Crankshafts transfer the motion of the piston to the wheels of a locomotive, and in this way heat was used to power trains and the industrial revolution. But after the piston has been driven to the top of the cylinder, it must obviously be returned to its original position so that the cycle can start over again, otherwise an engine would be like the *Titanic*, a one-time-only affair. To move the piston backwards requires relieving the pressure of the remaining steam in the cyclinder. So one might pass cold water around the cylinder in order to liquefy the vapor. The cold-water tubing or equivalent is naturally called a condenser and is the heat sink of a steam engine. Once the piston is back in its original position, the cycle is ready to begin anew.

The fact that you could get useful work out of burning coal and that the more coal you burned the more work you got led nineteenth-century scientists to conclude correctly that heat was a form of energy. The most peculiar aspect of all this was so evident that few people ever stopped to question it: you need a continuous supply of fuel to power a steam engine. When the piston returns to its original position and the hot steam condenses into cold water, it gives up its latent heat to the condenser. Why can't this heat be reused to drive the piston again and start the cycle all over, thus creating a perpetual motion machine? A theoretical answer was lacking, but centuries of failed attempts indicated that such a machine was not possible. And the natural question that ensued, just how nearly perpetual can you make a real engine, gave birth to thermodynamics.

The honor of "founder" of thermodynamics" is usually accorded Sadi Carnot, a Frenchman who died of cholera in Paris at the age of 36, just two months afer Evariste Galois, the subject of the last essay in this book, who avoided cholera long enough to find a bullet. Eight years earlier, in 1824, Carnot published his single memoir, a popular treatise entitled *Reflections on the Motive Power of Fire, and on Machines Fitted to Develop That Power.* He was continuing the work of his eminent father Lazare, an engineer, mathematician, and one of Napoleon's generals, who had concluded in 1803 that the most perfect or efficient engines were those that minimized friction between the parts, undue shocks, and rattle. Since all real machines are subject to such ills, Lazare drew the bold inference that a perpetual motion machine was impossible.

In *Reflections* Sadi used analogous reasoning and concluded that the most efficient engines were those that also transferred as little heat as possible between the parts. He devised an idealized engine in which the heat source and heat sink were entirely separate; consequently no heat could be exchanged between them. You must understand that the Carnot engine, as it is now called, was an idealization, a thought experiment. You cannot actually build one. It represents the theoretical limit, the best you can do. The best is not good enough. Using his engine, Carnot writes, "The impossibility of making the caloric produce a greater quantity of motive power than that which we obtained from it by our first series of operations, is now easily proved."[1] In other words, this most efficient of efficient engines is not a perpetual motion machine. You can easily calculate the efficiency of a Carnot engine and you find, in fact, that the efficiency is always less than or equal to one, meaning that the amount of energy you get out as useful work is always less than or equal to the amount of energy you have put in as heat.

I said that the Carnot engine was an idealization. Actually it is a double idealization. You can get an efficiency of one out of it but only when the temperature of the sink equals absolute zero. This is not very realistic. Also, Carnot assumed the engine to work in what physicists term the "quasi-static" limit, which means that the amount of time required to complete one cycle is infinite. This is not very realistic either. Any living engine has a lower efficiency than the Carnot engine, which itself has perfect efficiency only when the heat sink is at absolute zero.

Carnot's *Reflections* seem to have been shunted to the Oblivion File for some years and only the work of his colleague Emile Claperyon kept knowledge of them alive. Undoubtedly through Claperyon, Carnot's results came to the attention of Lord Kelvin and Rudolf Clausius, who were largely responsible for putting thermodynamics on a modern footing. Kelvin rejected Carnot's proof of "the impossibility of making the caloric produce a greater quantity of motive power . . ." as based on unsatisfactory assumptions. But the conclusion was so obviously true that he accepted it as a fundamental axiom. In Kelvin's words of 1852: "It is impossible for a self-acting machine, unaided by any external agency, to convey heat from one body to another at a higher temperature."[2] Or, as Clausius put it two years later: "Heat cannot by itself pass from a colder to a warmer body."[3] These are the first expressions of the second law of thermodynamics. They state that a perpetual motion machine is impossible. If a machine could spontaneously transfer heat from a cold body to a hot one, then the excess heat could be transformed into mechanical work via a steam engine. The extraction of heat from the cold body would cause it to become colder still, thus allowing (in this scenario) yet more heat to be transferred to the hot body. The engine would produce more work, still more heat would be extracted . . . and the sea would become infinitely salty. Such a situation struck Kelvin as patently absurd, and for good reason. Thus the second law.

An important point has been overlooked. Central to Carnot's reasoning was the idea that a perfect engine should be *reversible*; you can run Carnot's engine backwards and the imaginary piston ends up in exactly the same position as it started, and exactly as much heat would be absorbed as was originally released. But this perfect reversibility takes place only in the quasi-static limit already mentioned. In any realistic situation, you cannot get back to your exact starting point. We all know this to be true from everyday observation: the coal is burned, produces steam, and is reduced to embers. We never see fire reforge embers into usable coal. Something has been lost in any realistic, *irreversible* process.

This notion was made more precise by Clausius in the fourth of his famous memoirs on thermodynamics, where he states his version of the second law. There, assuming this principle to be true, he introduces a quantity which always grows regardless of the direction in which an irreversible engine operates. In his somewhat archaic language, ''The algebraic sum of all transformations occurring in an [irreversible] cyclical process can only be positive.''[4] The climax occurs in the ninth memoir of 1865 when the "transformation" receives a name:

> We might call S the *transformational content* of the body, just as we termed the magnitude U its *thermal and ergonal content*. But as I hold it better to borrow terms for important magnitudes from the ancient languages, so that they may be adopted unchanged in all the modern languages, I propose to call the magnitude S the *entropy* of the body, from the Greek word τροπή, *transformation*. I have intentionally formed the word *entropy* so as to be as similar as possible to the word *energy*; for the two magnitudes to be denoted by these words are so nearly allied in their physical meanings, that a certain similarity in designation appears to be desirable.[5]

Thus the increase in entropy follows directly from the second law, from the axiom that heat cannot flow from a cold body to a hot body. In fact, the two statements are equivalent and today we often see the second law stated as follows:

The entropy of an isolated system never decreases.

Understanding that the universe is the ultimate isolated system, Clausius leapt from the engines which generated his law to the cosmos and ended his ninth memoir with the following statement of the first and second laws of thermodynamics:

1. The energy of the universe is constant.
2. The entropy of the universe tends to a maximum.[6]*

Clausius tells us what entropy does—increases—but what exactly *is* this legendary quantity? To reduce some of the mystery, I present a formula that is valid for quasi-static processes in systems that receive heat from a reservoir: $ds = dQ/T$. Here, dQ is the amount of heat that has been extracted from the source, T is the temperature of the system, and dS is the resultant change in entropy. Since heat, as I've said, is a form of energy, we see that entropy looks very much like energy except that it is divided by a temperature. This is what Clausius meant when he said that "the two magnitudes . . . are so nearly allied in their physical meanings." In fact, entropy is the measure of the energy extracted from a heat source in an irreversible process that cannot be converted into mechanical work. If it were not for the T in the denominator, you could say that it is exactly the energy extracted from a heat source that cannot be converted into mechanical work. This is the no frills, orginal, nonmetaphorical entropy as it was conceived over a century ago.

It is important to bear in mind that the increase of entropy in thermodynamics is axiomatic. You cannot in any sense derive laws of nature; the best you could hope to do would be to show that one law of nature is based on another, which is then in turn postulated to be true. Such an attempt will be made in the next section. But as far as thermodynamics is concerned, the second

* In light of general relativity, Clausius's application of the first and second laws to the universe as a whole is not entirely accurate.

law is fundamental, logically equivalent to the statement that heat cannot spontaneously flow from a cold body to a hot body or that a perpetual motion machine is impossible.*

The profound and wonderful thing about the second law is that it gives an arrow of time in nature: the direction in which entropy increases is called "the future." Coal burns and is reduced to ashes, entropy is increased. This process occurs in only one direction—which we term "forward." The reduction of entropy and the phoenix rising from the ashes are seen only in legend. Why the second law should distinguish between past and future while all the other laws of nature do not is perhaps the greatest mystery in physics.

3. THE ENTROPY OF STATISTICAL MECHANICS

In an essay, "The Necessity of Art in Education," the well-known artist Jacob Landau speaks of " 'the paranoid style in American literature,' a style which, like an inversion of the ecological perspective, sees everything connected, but in sinister ways, and heading for chaos and entropy."[7]

The metaphor of entropy as cultural or artistic chaos is not new. When it first gained popular currency I cannot say, but the understanding that the growth of entropy represents an increase in disorder is due to the great Austrian physicist Ludwig Boltzmann (1844–1906) who, along with Maxwell, is regarded as the founder of statistical mechanics.

Statistical mechanics arose from the realization that the world is not as simple as fundamentalists would have us believe. A fundamentalist counts the number of generations in the Old Testament and arrives at the conclusion that the world was created in 4004 B.C., a celebrated result. Of course, to most of us, such a

---

* Perpetual motion machines are often divided into two classes. A perpetual motion machine of the first kind violates the first law of thermodynamics (conservation of energy), while a perpetual motion machine of the second kind violates the second law of thermodynamics. In this article we are only concerned with the second type.

calculation is frought with uncertainties. How can you be sure you counted all the generations between Adam and the birth of Christ? Methuselah may have lived 969 years, but what do you do after Noah, when the counting gets rough? Assign an average of thirty years per generation (Genesis, Chapter 11)?

The situation in physics is much the same. According to Newton's theory of mechanics, the motion of every particle in the universe can be described *exactly* by Newton's laws. Newtonian mechanics is therefore deterministic: give me the position and velocity of all the particles in the universe, and I will tell you exactly where each particle will be at any other time. A good example is the solar system itself. If I specify the position and velocities of all the planets at a given instant, I will be able to predict with exact certainty where all the planets will be in one year or where they were in 4004 B.C. You might call Newtonian mechanics the fundamentalist view of physics.

Newtonian mechanics also contains another peculiarity: like the Carnot cycle, it is *time-reversible*. This is often said to mean that if I ran the solar system forward for an hour, then backwards for an hour, the planets would all end up exactly where they started. The statement is true but "time-reversible" has a more general meaning with regard to Newtonian mechanics: if you watch a film of the planets orbiting the sun clockwise, you nevertheless have no right to say that clockwise means forward in time. For all you know the film might have been shot backwards to begin with so that in fact counterclockwise is really forward. Newton's equations are equally good in either direction, much like the White Queen's memory, which remembers past or future equally well. We call such equations time-symmetric.

Not only are the equations of Newtonian mechanics time-symmetric, but those of electrodynamics, relativity, and quantum mechanics are also. That the equations of the most fundamental physical theories are time-symmetric is a deep peculiarity, for we all know real memories only look backwards and phoenixes do not rise from the ashes any more than shattered cups spontaneously reassemble or mushroom clouds implode into undetonated atomic bombs.

The profound mystery I mentioned at the close of Section 2 can now be restated: If the world is truly based on Newtonian mechanics and on its successors, which are all time-reversible, how can the world we observe be irreversible? Some aspects of the real world, namely its irreversibility, are more like thermodynamics than Newtonian mechanics. But this brings us immediately to the central paradox. Thermodynamics, based on the time-asymmetric axiom of the second law, is then incompatible with time-symmetric Newtonian mechanics. Two great pillars of physics stand in apparent contradiction.

Newtonian mechanics claims to govern the behavior of individual particles: atoms, molecules, billiard balls. Thermodynamics deals with large aggregations of particles, such as lumps of coal and engines full of steam; it also talks about the flow of heat—a "fluid" that results from the action of many particles. So it is natural to assume, contrary to the formulations of Kelvin and Clausius, that the second law is not axiomatic. Instead, reversible Newtonian mechanics is the more fundamental theory, and when a large number of particles is brought into play, thermodynamics, the second law, and irreversibility somehow emerge. From this point of view, Newton's laws are *microscopic* and the second law is *macroscopic*.

This was in fact the philosophy of Boltzmann, who attempted to resolve the schism between mechanics and thermodynamics with a decidedly nonfundamentalist approach. A bread box contains roughly 10^{23} molecules of air. Just as it is impossible to know exactly how many generations elapsed between Adam and Jesus, it is equally impossible to know the trajectories of each of the 10^{23} molecules in a bread box. Boltzmann therefore introduced statistics (hence the term statistical mechanics). He averaged velocities, averaged positions, and, using these averages, calculated the most probable behavior of a gas. An analogous method would be to assign an average of thirty years per generation in the Old Testament and thereby estimate the time elapsed from Adam to the writing of the Bible. The average may not represent any particular generation but one would hope that the result was reasonable.

Using the statistical approach, Boltzmann published a famous paper in 1872 where he claimed to show that a certain quantity, called H, always decreased. H was identified with the negative of another quantity, S, which consequently always increased. Boltzmann's formula for S was written as

$$S = k \sum_i p_i ln p_i.$$

In this equation k is just a number called *Boltzmann's constant* and p_i refers to a certain probability which I will explain in more detail later. The symbol Σ_i indicates that we are to sum over all probabilities p_i. Because the S of Boltzmann's H-theorem is always increasing, one is immediately tempted to identify it with the S of thermodynamics. But at first glance the two formulas don't look very similar, the S of thermodynamics containing a temperature T and the S of statistical mechanics containing some probabilities p_i. Nevertheless, with a sufficient amount of work one can show that the two S's represent the same thing. So at first glance Boltzmann has pulled off a triumphant sleight-of-hand. Starting from Newtonian mechanics and introducing statistical procedures he has managed to derive thermodynamic entropy.

It is for this reason that the general increase of entropy is often said to be due to our ignorance of the individual trajectories of particles. The particles are too numerous to follow individually, so we perform some averaging (called "coarse-graining" in physics) and—*voila!*—entropy. But surely this is very strange. From a time-symmetric theory he has managed to produce a time-asymmetric result. Sleight of hand.

A simple model devised by Dilip Kondepudi and myself contrasts the Newtonian and statistical-mechanics view of entropy and illustrates the difficulties inherent in Boltzmann's approach. It takes as its point of departure the famous "Ehrenfest urn model," which contains too many balls to be immediately comprehensible.

Consider two labeled urns, A and B, and four numbered balls, 1, 2, 3, 4. The balls are moving one at a time between the two urns. Now imagine two demons who are watching the balls move; Isaac, who can read the numbers on the balls; and Ludwig, who is so nearsighted that he cannot. At any given instant the balls will be found divided between urns A and B in one of the sixteen possible ways shown in Table 4.1.

The first three columns refer to Isaac's view of the situation. Since he can read all the numbers on the balls, he sees sixteen distinct states, labeled in the first column as 1–16. The last three columns refer to Ludwig's view. This requires a bit more explanation. Because he cannot distinguish the numbers on the balls, states 2–5, for example, all appear to him to be exactly the same— merely one ball in urn A and three in urn B. To him they are all one state which he labels (1,3) in the last column. Similarly, what to Isaac are the distinct states 6–11 appears to Ludwig as the single state (2,2).

Suppose that the balls move once each second and that Ludwig watches the transfer for an hour, or any other sufficiently long

TABLE 4.1

State	Isaac		Ludwig		State
1	____	1234	____	0000	(0,4)
2	1	234	o	000	(1,3)
3	2	134	o	000	(1,3)
4	3	124	o	000	(1,3)
5	4	123	o	000	(1,3)
6	12	34	oo	oo	(2,2)
7	13	24	oo	oo	(2,2)
8	14	23	oo	oo	(2,2)
9	34	12	oo	oo	(2,2)
10	24	13	oo	oo	(2,2)
11	23	14	oo	oo	(2,2)
12	234	1	000	o	(3,1)
13	134	2	000	o	(3,1)
14	124	3	000	o	(3,1)
15	123	4	000	o	(3,1)
16	1234	____	0000	____	(4,0)

period. Being inclined toward statistical mechanics he will assume the balls are moving at random. What does he see? First, he will notice that the state (0,4) with no balls in A and all four in B does not occur very often. From the table you can see that it should take place about $\frac{1}{16}$ of the time, since there is only one way out of sixteen possible ways to make this configuration. On the other hand, the balls will end up in the state (1,3) about $\frac{1}{4}$ of the time and in the state (2,2) very nearly $\frac{6}{16}$ or $\frac{3}{8}$ of the time. To anticipate slightly, we will refer to the highly improbable configurations (0,4) and (4,0) as low entropy states and the most probable (2,2) configuration as the high entropy state. The configurations (1,3) and (3,1) are intermediate entropy states.

Ludwig now makes several observations. If the balls are in state (0,4), then after the next move the only possible outcome is the state (1,3). So the *transition probability* from (0,4) to (1,3) is 1 (meaning 100% likely). Similarly, if the balls begin in state (1,3) and one ball moves at random, then there is a 25% probability that you will end up back in state (0,4) but a 75% probability that you will end up in the state (2,2). So the transition probability from the intermediate entropy state (1,3) to the low entropy state (0,4) is only $\frac{1}{4}$, but the transition probability to the high entropy state (2,2) is $\frac{3}{4}$. Note that *the transition probability from (0,4) to (1,3) is not the same as from (1,3) to (0,4).*

Ludwig computes all the transition probabilities between the various states and draws the following diagram:

$$
(0,4) \quad \overset{\textstyle \frac{1}{4}}{\underset{\textstyle 1}{\rightleftarrows}} \quad (1,3) \quad \overset{\textstyle \frac{1}{2}}{\underset{\textstyle \frac{3}{4}}{\rightleftarrows}} \quad (2,2)
$$

——— entropy increases ———→

The demon looks at his results and exclaims, "Aha! The probability of going from left to right is always higher than from right to left. Nature tends to go from the least probable to the most probable. Entropy increases!" This increase in entropy defines the statistical-mechanics arrow of time.

But now the demon Isaac objects: "Wait a minute. I see sixteen distinct states. No state is any more 'probable' than another. Your increase in entropy only came about because you first averaged distinct states into indistinct states and then assumed the balls are moving at random. But my world is deterministic. Balls do not move at random. If a ball moves and changes state 11 to state 12, that is because it has been preordained in the past. Furthermore, my world is time-reversible so if I reverse the clock, state 12 goes back to state 11. There are no 'unequal transition probabilities' between configurations. Your so-called entropy arose only because you introduced statistics. I reject this nearsighted view of nature. Entropy does not exist.''

This model demonstrates essentially what Boltzmann did to get the entropy of statistical mechanics. Using his formula above we can actually show that, as claimed, the state of highest probability (2,2) is also the highest entropy state. It is important to realize that the p_i in the formula do not refer to the probability of the (2,2) occurring—which is $\frac{3}{8}$—but refer to the probability that a given ball will be in jug A or B. For (2,2) this is obviously 50-50, or $\frac{1}{2}$ for each urn. The formula gives

$$S(2,2) = -k[\frac{1}{2}ln\frac{1}{2} + \frac{1}{2}ln\frac{1}{2}] = .693k.$$

For the state (1,3) the probability is 25% that a ball will be in jug A and .75 that a ball will be in jug B, so we have

$$S(1,3) = -k[\frac{1}{4}ln\frac{1}{4} + \frac{3}{4}ln\frac{3}{4}] = .562k.$$

And for (0,4) we get

$$S(0,4) = -k[0ln0 + 1ln1] = 0.$$

So the most probable state (2,2) is also the one with the highest entropy and the least probable state (0,4) has the lowest entropy.

Onward. Imagine a room in which all the gas molecules are shoved into one corner, analogous to the situation where all four balls were in one urn. This configuration is called highly asymmetric or ordered—asymmetric because all the gas molecules in one corner or all the balls in one urn is indeed an extremely asymmetric case (two balls in each urn would be symmetric); ordered because it rarely occurs. The most probable random outcome of scattering molecules or balls or clothing is for them to end up lying disordered uniformly all over the room or distributed equally in the two urns. Barring a freak accident, the ordered state (balls in one urn, molecules in one corner, clothes in bureau) usually comes about only with the expenditure of considerable energy. Boltzmann's results show that, regardless of the initial conditions, with the passage of time the most likely configuration is the symmetric state—clothes scattered across the room, (2,2)—which is the state of maximum entropy and disorder. This state is also known as the state of equilibrium, since once you attain complete disorder things tend to stay that way. Here then is the origin of Jacob Landau's metaphor that high entropy is equivalent to chaos.

The purpose of our model was not only to demonstrate Boltzmann's approach but the defects in it. From the demon Isaac's point of view, the result is entirely subjective—an artifact of Ludwig's nearsightedness, or more technically the introduction of statistics. The subjective nature of Boltzmann's interpretation is very unsatisfying, for it means that if Ludwig's blind grandfather happened by, he might not even see two distinct urns and therefore come to yet completely different conclusions about the system. Perhaps he would deny the existence of motion itself.

There is yet a further difficulty here, a subtle point and one infrequently mentioned. Recall that the transition probability from state (0,4) to (1,3) was 100% but the reverse transition probability from (1,3) to (0,4) was only 25%. For this reason Ludwig concluded that nature's irreversibility stemmed from its tendency to go from least probable to most probable. However, if you sit in state (0,4), not only is the probability 100% that the next state will be (1,3) but the probability is also 100% that the *previous* state was (1,3). [The only possible way to get to (0,4) is from (1,3).] Similarly, if you are in (1,3), we found the probability to be 75% that the next move will bring (2,2). But the probability is also 75% that the *previous* state was (2,2). Therefore the probability that the next state will be of higher entropy is exactly the same as the probability that the previous state was of higher entropy. So we see that even with the introduction of statistics, the system is time-symmetric *unless we use only the notion of probability to predict in one direction, called the future.* *

We unhappily arrive at the conclusion that the arrow of time or the increase in entropy in statistical mechanics depends on two necessary assumptions: (1) that motions are random; (2) that we introduce averages. These assumptions are, however, not sufficient of themselves and a third must be added: (3) that we use probabilities to calculate in only one direction. In this way, the increase of entropy is once again found not to be a result of mechanics (or even statistics) but of the ad hoc assumption that probability only predicts the future, not the past. This point seems to have been first recognized by Gibbs.[8]

To sum up, the entropy of statistical mechanics is equivalent to that of thermodynamics but, contrary to Boltzmann's expectations, the equivalence does not follow directly from mechanics, even when statistics are introduced. A hidden assumption about the nature of probability is also required, and thus the incompatibility of thermodynamics and Newtonian mechanics remains unresolved.

4. THE ENTROPY OF INFORMATION THEORY AND COMPUTATION

Writing on that champion consumer of mass information, the television watcher, Jacob Landau notes:

For the *watcher* the world is small—19 inches diagonally, lacking in grandeur, in fact quite piecemeal and tawdry. Because he is on the surface on which events, quite meaningless in their interconnections, are projected, he feels himself to be larger than life, he is egocentric, narcissistic. Electric integrationism drives out linear or causal understanding, resulting in explosive anxiety, apocalyptic violence, atomization, totalism—or numbing, narcosis. The intuition that media

---

* Although this point was mentioned briefly in the text of the *Discover* article, it unfortunately was not made in the exposition of the model (box, pp. 76–77, February 1987).

happenings may have obsoleted all other art forms is founded
in the awareness that a quantum leap into total unreality has
taken place, that all communication is now art, so nothing is
art. Reality inverts to illusion, and art to anti-art—a prelude
to maximum entropy and/or the big bang.

The consequences for education are staggering. Increased
emphasis on the cognitive "basics," redoubled efforts to be
logical or rational, will only increase the entropy and paranoia.[9]

We now make a quantum leap back to reality. Entropy and information—or more accurately, the lack of it—have become
synonymous, like "Princeton" and "traffic." The metaphor has
its origins in communication or information theory (two inter-
changeable terms), which was developed largely by Claude Shannon at Bell Labs in the late 1940s.

Shannon, like his predecessors, was concerned with the most
efficient way of encoding and transmitting messages across, say,
a wire or any other channel. In his famous 1948 paper, "A Mathematical Theory of Communication,"[10] he used the term "entropy" to describe the number of binary digits (bits) required to
transmit a given message. From that moment on the confusion
between the entropy of information theory and the entropy of
physics was complete. Since then, many attempts have been made
to marry the two concepts but, as John R. Pierce writes in *An
Introduction to Information Theory*, they "have been more interesting than fruitful."[11]

One problem is that Shannon deliberately defined communication theory entropy by a formula designed to look as much like the
formula of statistical mechanics as possible:

$$S_c = \sum_i p_i \log_2 p_i \text{ bits},$$

Where S_c denotes the "entropy of communication," $\log_2$ indicates
the traditional use of base-two logarithms, and p_i is the probability
of choosing a given character or word from all available characters
or words.

As an example, suppose we wish to transmit a message to a friend, telling him whether a flipped coin has turned up heads or tails. We have two available words, "heads" and "tails," or equivalently zero and one in bits. For a random toss, the probability of choosing a head is of course ½ and the same is true for tails. Thus p(heads) = ½, p(tails) = ½, and the formula gives

$$S_c = -[½\log_2(½) + ½\log_2(½)]$$
$$= 1 \text{ bit per toss.}$$

Not surprisingly, we need one bit—either a 0 or a 1—to transmit the result of a coin toss.

If, on the other hand, the coin were weighted so that the probability of getting heads is zero and the probability of getting tails is one, Shannon's formula gives us $S_c = 0$. The entropy has been reduced to zero. Intuitively this makes sense; you don't need any bits to transmit the result of this experiment because you know the result in advance. The same holds true for an honest coin once it has been flipped; when the probability of, say, tails is known to be 1, $S_c = 0$. [Note also the similarity to the urn model, in particular the state (0,4) where we know with certainty the position of each ball.]

We see that communication-theory entropy is a measure of the uncertainty of producing a given message from among the available options. For an honest coin before it is flipped you need one bit because the outcome is uncertain. For a weighted coin or an honest coin after it has been flipped, the outcome is certain and you don't need any bits. The more uncertain the outcome, the greater the entropy and the greater the number of bits you need to transmit a message. The number of bits you need to transmit the message and reduce the entropy to zero is termed *information*. (In fact the change in entropy is exactly equal to the negative of the information.) Here then is the origin of the metaphor of entropy as lack of information. It is important to distinguish between "information," in this technical sense, and information as we normally use it in the sense of *knowledge*. Paul Davies has written: "Everyday experience indicates that information only *increases* with time. Our own memories grow as we do, public libraries accumulate books, the moon accumulates craters from meteoric impacts.''[12] Davies is strictly correct, but misleading because he does not carefully distinguish information from knowledge. If I generated a 900-page book by randomly picking letters from a giant urn, it would take about as many bits of information to transmit the characters contained in that book as it would to transmit the characters in *The Brothers Karamazov*. But because the sequence of letters in my book would be entirely random, the knowledge content would be zero. If libraries were filled with such randomly generated books, the information content would indeed increase with time but knowledge would not. One wonders whether this is actually the case.

From this discussion we see that the considerations leading to the entropy of information theory differ considerably from those that led to thermodynamics and statistical mechanics. In Shannon's equation there is no Boltzmann's constant (though see below); his entropy apparently has nothing to do with temperature nor with gas molecules. There is no statement of the second law here. Yet one feels that the statistical-mechanics entropy and the information-theory entropy play such similar roles that there ought to be a connection. Whether we speak of the probability of tossing a head or the probability of finding a ball in urn A, the principle appears to be the same. But what message can we extract from a box of gas? The connection is made with the realization that it takes energy to transmit information.

The story begins in 1867 with a famous thought experiment by Maxwell that he mentioned in an 1870 letter to his friend Lord Rayleigh. He was objecting to the second law of thermodynamics:

> For if there is any truth in the dynamical theory of gases, the different molecules in a gas of uniform temperature are moving with very different velocities. Put such a gas into a vessel with two compartments [A and B] and make a small hole in the AB wall about the right size to let one molecule through. Provide a lid or stopper for this hole and appoint a doorkeeper very intelligent and exceedingly quick, with microscopic eyes, but still an essentially finite being. Whenever

he sees a molecule of great velocity coming against the door from A into B he is to let it through, but if the molecule happens to be going slow, he is to keep the door shut. He is also to let slow molecules pass from B to A but not fast ones. . . .

In this way the temperature of B may be raised and that of A lowered without any expenditure of work, but only by the intelligent action of a mere guiding agent. . . .

Moral. The 2nd law of thermodynamics has the same degree of truth as the statement that if you throw a tumblerful of water into the sea, you cannot get the same tumblerful of water out again.[13]

This "exceedingly quick being" was dubbed by Rayleigh "Maxwell's Demon" and has plagued physics ever since. To recapitulate, the Demon, acting like "the pointsman on a railway with perfectly acting switches who should send the express along one line and the goods along another," separates the molecules in Maxwell's box into slow (cold) molecules on one side of the partition and fast (hot) molecules on the other. Since no energy has been expended in the process, he has created a perpetual motion machine, for now the temperature difference in the two sides of the box can be exploited to run a steam engine.

Many resolutions have been proposed to Maxwell's Demon. The most celebrated of these, usually attributed to Szilard, is actually due to Demers and Brillouin.[14] What Maxwell did not know (and could not have known) in 1867 was that any object, including a box of gas, at a finite temperature emits and absorbs radiation called *black body radiation*. If you heat an iron bar in a furnace until it glows red, you are seeing the black body radiation characteristic of about 5000°. The important point is that you are seeing the radiation, not the iron; any object heated to 5000° will glow the same color. Maxwell's Demon, staring into a box of gas at 5000° would see only a red glow. He could not see the individual molecules because they would be absorbing and emitting only this red light. Since he could not see the molecules, he would not know when to open the door and would fail in his mission to violate the second law.

To get around this obstacle, the Demon would have to shine a light of a different color onto the molecules and record this reflection in his eye. This reflection constitutes one bit of information. However, the light bouncing off the molecule is absorbed by the Demon's eye and, since a molecule contains heat energy, the absorption represents a slight increase of the Demon's own entropy. And because the Demon is at 5000° as well, this entropy increase is transferred to the the entire system. Brillouin was able to show that the increae is just enough to prevent the Demon from violating the second law. As Dennis Gabor, inventor of the hologram, put it in 1951, "We cannot get anything for nothing, not even an observation."

Here then is the connection between the entropy of statistical mechanics and the entropy of information theory: by knowing "which urn" the molecule is in, the Demon has decreased the uncertainty of its position and consequently decreased the stat-mech entropy of the gas. Maxwell's desires are fulfilled. But the observation requires the extraction of one bit of information. The price for this bit of information is the increase in stat-mech entropy of the Demon himself. It is equal to $kln2$ and now we see a formula reminiscent of statistical mechanics. This point is more clearly illustrated by Szilard's refutation of the Demon, which I include largely because I believe the argument to be somewhat incorrect.

Consider the following sort of steam engine: A piston is housed in a cylinder such that it is free to slide in either direction. To each side of the piston is attached a rope so that if the piston moves to the right, the left-hand rope is pulled and this motion can be used to, say, hoist a rock. If the piston moves to the left, the other rope is tugged and a rock on the right-hand side is lifted. This steam engine will work using one molecule of steam. We center the piston and open a little hatch in its center, allowing the molecule to wander freely between the two halves. Once we know that the molecule is on, say, the right-hand side of the piston, we close the hatch and tie a rock to the right-hand rope. Now, this molecule will knock around in the cylinder and repeatedly bang the piston, pushing it to the left, and in the process raise the rock. Just by knowing what side of the piston the molecule is on, we seem to have got some work.[15]

The catch, according to Szilard, is in the "just by knowing." To know where the molecule is requires one bit of information (left or right, 0 or 1). Szilard was able to show that to transmit this bit of information requires just enough energy ($kTln2$) to cancel out any gained by the work of the molecule on the piston. Once again, by reducing the uncertainty of the molecule's position and the stat-mech entropy of the system, we lose out elsewhere. In this case reducing the stat-mech entropy frees up energy available for work. But the amount of energy freed up for work turns out to be just equal to that needed to transmit the bit of information to the Demon. So he hasn't gained anything.

In this limited way, the entropy of information is connected with the entropy of statistical mechanics. The message was of a very peculiar kind—it specified the state of the system. A message in English doesn't specify the state of any system and its connection to statistical mechanics appears to be tenuous at best. And if I were to propose a "second law of information theory," that information-theory entropy always tends toward a maximum, I would be on very metaphorical ground indeed.*

But as I said, I believe Szilard's argument (though not its conclusion) to be flawed. It's fairly clear that his engine will work for at least one stroke regardless of whether we know what side the molecule is on. Tie a rock to both ropes. If the molecule is on the right-hand side, the right-hand rock will be lifted; if the molecule is on the left-hand side, the left-hand rock will be lifted. In either case we have gotten some work out of the beast. I leave it as an exercise for the reader to devise a correct refutation.[16]

Recently there has been considerable controversy over "logically reversible computers." These are idealized machines which are meant to perform the usual logical operations of computers without the expenditure of energy. Such machines—computa-

---

* Not least because information theory deals with the transmission of information between two systems but the second law of statistical mechanics and thermodynamics refers to a single isolated system.

tional analogs of Maxwell's Demon—have been extensively investigated by Rolf Landauer and Charles Bennett at IBM.[17] However, at least one such model, known as the "billiard ball computer," has been shown by Wojciech Zurek not to work without the usual increase of entropy already discussed. I suspect that the other models will eventually suffer the same fate.

5. ENTROPY AS EVOLUTION

The central thesis of the 1986 University of Chicago Press book *Evolution as Entropy* by Daniel Brooks and E. O. Wiley[18] is that the second law of thermodynamics is responsible for biological evolution. While any physicist would instantly admit that evolution must be compatible with the second law, it is a much greater leap to suppose that evolution is a necessary consequence of the second law, which is what Brooks and Wiley maintain.

I readily acknowledge that when I first encountered the book I had difficulty getting past the title. Evolution is a process. Entropy, as I've discussed it, is a quantity, similar to energy. Brooks and Wiley have equated a process to a quantity and to this day I don't know what they mean.

Judging from the dust-jacket endorsements of eminent persons, I am alone. Nevertheless, the authors are respected biologists and I thought the book deserved a serious glance between the covers. The glance turned into a lucubration and the results proved illuminating.

But before I take issue with the authors, I should first clarify the thesis. One's first reaction to the claim that the second law governs evolution might be to argue the contrary—that evolution stands in direct violation to the second law. The second law represents an increase in disorder and chaos, while evolution leads to ever more complex and highly ordered life forms. Evolution is therefore incompatible with thermodynamics. Creationists frequently make this argument in order to prove the impossibility of evolution. The argument is incorrect. The strict increase in entropy applies only to isolated systems. In physics one distinguishes between three types of systems: isolated, closed, and open. I list their definitions now, for the distinction will prove to be important:

1. *Isolated systems*: Completely disconnected from their surroundings; no exchange of energy or matter possible.
2. *Closed systems*: May exchange energy with their surroundings but not matter.
3. *Open systems*: May exchange energy and matter with their surroundings.

While isolated systems tend toward maximum entropy, it has been known for many years that the equilibrium state of closed and open systems is not necessarily one of maximum entropy. A closed system, for example, may reduce its internal entropy at the expense of increasing the entropy of its surroundings. This is in fact how life is maintained. So the creationists are wrong.

The authors of *Evolution as Entropy* go to the opposite extreme and argue that the increase of entropy actually causes evolution. Despite creationists, the metaphor of entropy as evolution is easy to see: the increase of entropy and evolution are both irreversible processes; therefore the two processes must be related. Presumably Brooks and Wiley (B&W) wish to put the metaphor on a firmer footing and consequently entitle their key Chapter 2 "Why Entropy?" I do not believe they ever answer their own question. At the beginning of this chapter they quote astronomer David Layzer: "But the Second Law is purely macroscopic; it has no microscopic counterpart. Hence, the directionality of the Second Law must be a consequence of auxiliary conditions."[19] "Auxiliary conditions" refer to what physicists usually call boundary conditions. Layzer means that if entropy is now increasing it must be because the system, for example the universe, was created with the lowest possible entropy. In this case we say that the initial boundary conditions of the universe were low-entropy conditions. B&W continue: "Prigogine and Stengers (1984) have recently reiterated the division of macroscopic systems into those whose behavior is caused by their boundary conditions, which they call *thermodynamics systems*, and those whose behavior is caused by their initial conditions, which they called *dynamic systems*. We can readily see that this dichotomy is an artificial one, at least for biological systems.''[20]

By quoting Layzer and then immediately passing to Prigogine and Stengers, B&W imply that Prigogine believes the second law to be macroscopic. On many occasions Prigogine has emphasized that this is not his position. For my *Discover* article on the topic he said: ''We are living at a very interesting moment. We can no longer say that irreversibility is the result of an approximation or lack of knowledge, and we must now decide what sort of microscopic laws lead to irreversibility.''[21]

Furthermore, initial conditions are merely a form of boundary conditions—conditions that pertain at the beginning of a problem. To distinguish between initial conditions and boundary conditions is artificial. I believe Prigogine would argue that thermodynamic systems (ones that demonstrate an increase in entropy) are those systems in which the initial conditions are impossible to determine exactly, whereas dynamic systems (ones that obey Newton's laws) are those systems in which the initial conditions can be exactly determined.

Arguing over interpretation of a third party is admittedly not the most fruitful endeavor. But in Chapter 2, B&W make a number of statements which are open to less subjective criticisms. They equate closed systems with isolated systems (''Type I systems are sometimes called *closed*, or isolated . . .''[22]) and then go on to describe the energy change of the system. However, another celebrated law of thermodynamics, namely the first, tells us that the energy change of an isolated system is zero. By definition. On the same page B&W discuss yet another thermodynamic quantity called the ''Gibbs free energy,'' usually denoted by G. In contrast to entropy, G characterizes the amount of energy available for nonmechanical work. B&W go on to state that spontaneous fluctuations in G will take place if the temperature changes. But by assumption fluctuations in G take place at constant temperature.[23]

The book is filled with these sorts of mistakes which will make any physicist highly skeptical of their conclusions. More to the point of this essay, the text is filled with metaphors and other vaguenesses which, though attractive to poets, are anathema to physicists:

> . . . we must reject theories in which ecological constructs ("niches," "adaptive zones") serve as analogues of quantum states. . . .[24]

> . . . we will assume that any macroscopic system that is reproducible enough to be described topologically has an entropy and its entropic behavior can be measured.[25]

> . . . how can we use Boltzmann's notions of statistical entropy and reject a quantum view of biology?[26]

While I can appreciate the use of "quantum state" as a metaphor for "ecological niche," I can also appreciate the use of "quantum state" as a metaphor for "rat race." Why not "the quantum view of commuter life"? The relationship of entropy to topology far exceeds the capacities of my simply connected mind.

Later in the book,[27] B&W discuss extensively the concepts of "cohesion," "cohesion entropy," and "entropy of a population." Cohesion, as I understand their use of the term, is a measure of how well a species or population "sticks together." Presumably as members of one population mate with members of another, the cohesion within each population goes down and the population entropy goes up. The formula B&W use to compute the cohesion entropy is exactly the one I discussed in Section 4 on information theory, and they express their answer in bits.

At this point one must ask whether the thread connecting entropy with thermodynamics still exists, albeit frayed, or if it has snapped completely. If "population entropy" has anything to do with thermodynamics, B&W must then be able to tell me what the temperature of a population is. If they are using entropy in its statistical-mechanics sense, then I want to know what they mean when they say that entropy measures the "number of microstates accessible to the population." Quantum states? I'd hesitate to begin counting. If they mean entropy in the communication-theory sense, I then want to know what their bits measure and what they mean by the statement, "Our concept of information is not completely compatible with traditional views of information stemming from communication theory."[28] And if the entropy of information theory itself is only tentatively related to that of thermodynamics, what do we have here? A second-order metaphor?

Richard Feynman, in his memoirs, talks about "cargo cult science."[29] He takes his metaphor from the South Sea islanders who, during World War II, saw airplanes bring to their islands many and useful supplies. Since the war the planes have disappeared but, in an attempt to lure them back, the natives have made runways, built fires next to them, assigned controllers to wear wooden headphones with bamboo antennas. They have the form right; what is lacking is the content, and their efforts have been unsuccessful. Toward the end of *Evolution as Entropy*, B&W write: "In physics, relativity theory subsumed Newtonian mechanics. In postulating that our theory represents a kind of relativity theory itself, we find ourselves in an analogous position."[30] Analogy is the operative word in *Evolution as Entropy*. The work represents less a theory than a metaphor, a grand mixing of metaphors. Time will tell whether their metaphors bear any content. But the one analogy I am certain will not hold is the one I just quoted.

6. ENTROPY À LA RIFKIN

Bach violin partitas and orchestral suites almost invariably begin with the most difficult movements and end with the simplest. Brahms follows the same pattern in his sextets and orchestral serenades and Bartók uses the identical strategy in his *Music for Strings Percussion and Celesta*, which starts off with a monstrous fugue and effervesces into a quirky folk tune. His unaccompanied violin sonata also violates the second law by passing from the complex to the straightforward. The plan has survived, for it is a sound one; after a long partita or sonata, both performer and listener are losing endurance and concentration and deserve a reprieve. I therefore end this essay with a joke.

In 1980 Jeremy Rifkin published a book entitled *Entropy: A New World View*.[31] In 1983 he published *Algeny: A New Word—*

A New World, whose first chapter is subtitled "A New Metaphor for the Coming Age."[32] To overturn the world order twice in the space of three years is a feat not even Einstein can lay claim to, yet judging from the dust-jacket testimonials of eminent persons, Rifkin has done just that.

In a rational world, *Entropy* would have been published (or perhaps not), dismissed with a horselaugh, and forgotten. However, the book became a bestseller, and years later it is still cited; moreover, Rifkin uses the same type of arguments in his more recent campaign against genetic engineering, and at least one Texas graduate has begun "entropy education courses." Rifkin's brand of pseudoscience (and make no mistake, it is pseudoscience) is much more difficult for the public to perceive as tomfoolery than the usual anesthetics of scientology and astrology because Rifkin claims not to be attacking established science, but to be using it to attack the establishment.

These tactics interest me more than his thesis (which is trivial), so let me start with tactics. A good example is the following early passage, where the anti-science tone of the book is made clear:

> Strangely enough, while scientists have anguished over the proper meanings of [the first and second laws] for longer than anyone cares to remember, they were already well established in the everyday folklore of just about every culture on earth. How many times have we heard the phrase "You can't get something for nothing" or "It does no good to cry over spilt milk" or "You can't beat the system." If you are familiar with these phrases and have seen them verified in your own everyday experience, then you know about the first and second laws of thermodynamics.[33]

Scientists frequently lack wisdom and insight. On the other hand, if the common man's faculties in this regard are so great, we might wonder why attempts to build perpetual motion machines persist, why Americans switch to large automobiles every time the price of gasoline drops, why they haven't solved the energy—or entropy—crisis, and why Rifkin needs to write his book.

Not long after, the full-scale attack on scientists is launched:

Starting from maximum entropy or a total equilibrium state of uniform energy, Maxwell proposed to reverse the entropy process without any outside energy being used; this would have violated the second law. First it's obvious that in the real world we'd never be able to produce such a demon. But just to humor Maxwell, let's assume an appropriate demon could be found and that it would be willing to take on the job. Could it perform its work without violating the second law? Stanley Angrist and Loren Helper, writing in *Texas Quarterly*, put the demon to the test and discovered that it could not get around the Entropy Law:

> [Maxwell] supposed that his demon would be able to sense the velocity (speed and direction) of individual molecules and then act accordingly. . . . As the demon peers into either side of the isolated enclosure at uniform temperature, the uniformity of radiation throughout does not permit him to see anything. The sameness in the enclosure would allow him to perceive the thermal radiation and its fluctuations, but he would never see the molecules. . . . We conclude that the demon needs his own supply of light to disturb the radiation equilibrium within the enclosure. . . .

About the only thing this whole exercise proves is that "we cannot get anything for nothing, not even an observation."

Maxwell's attempt to challenge the Entropy Law is worth remembering. It is, more than anything else, a reflection of the hardheaded refusal of the scientific community to acknowledge the full implications of what the Entropy Law means for science, philosophy, and life on this planet.[34]

Recalling Section 4, you will have recognized this refutation of Maxwell's demon as not originating in 1967 with the *Texas Quarterly* but in 1951 with Brillouin, a physicist. You will also have recognized the unattributed "we cannot get anything for nothing, not even an observation" as coming from Dennis Gabor, another physicist. Conjecture and refutation is the dialectic of science. Straw men and false attribution is the dialectic of Rifkin.

Now that Rifkin has dismissed hardheaded physicists by using an argument they devised and which he professes to believe, he shows further acrobatic skill by embracing a principle which is totally incompatible with the one he has just put forth. Rifkin quoting his master, economist Nicholas Georgescu-Roegen:

> Every farmer understands that, even with recycling and constant sunshine, it's impossible to grow the same amount of grass on the same spot year after year in perpetuity. Every blade of grass grown today means one less blade of grass that can be grown in the future on the same spot. The recognition of this fact is incorporated in the fourth law of thermodynamics, first advanced by Nicholas Georgescu-Roegen: "In a closed system, the material entropy must ultimately reach a maximum."[35]

Far be it for me to argue about blades of grass, but I suspect the press run of *Entropy* destroyed more potential saplings than bad farming. At any rate, if garbage (excuse my replacement for the highbrow "material entropy") must reach a maximum in a closed system, then the total entropy (including the "material" sort) could still increase while the usual thermodynamic entropy of previous mundane deliberations could decrease. But then we could build perpetual motion machines and solve the energy crisis. The new fourth law in contradiction to the old second!

But extending the frontiers of entropy beyond known physics is not good enough for Rifkin and so he strides forth to biology. Genetic engineers beware:

> In an effort to ignore the Entropy Law, the experts will attempt to convince the rest of us that with a renewable energy base we will never run out of resources, and that growth will go on forever. In the short run, new genetic technologies, like recombinant DNA, might greatly increase the matter-energy flowing through the system, just as the first industrial transformation did with nonrenewables. For a time at least, it may well appear that we have overcome the fixed limits of the earth's ecosystem. That time span will be short-lived.[36]

A flag of logic to be raised here: why can't genetic engineering be used to reduce the energy flow? But why worry about such details when you can use the Entropy Metaphor to escape the bounds of science altogether:

> Adherents of the Eastern religions—and especially the Buddhists—have long understood the value of minimizing energy flow-through. The practice of meditation is designed to slow down the wasteful expenditure of energy. The state of Nirvana or truth is reached when the individual is expending the least energy necesssary to support his outward physical survival. The Eastern religions have long claimed that unnecessary dissipation of personal energy only adds to the disorder and confusion of the world.[37]

And on to the infinite:

> Love is not antientropic, as some would like to believe. If love were antientropic, it would be a force in opposition to becoming, for the entropic flow and becoming go hand in hand. Rather love is an act of supreme commitment to the unfolding process. That is why the highest form of love is self-sacrifice—the willingness to go without, even to give one's own life, if necessary to foster life itself.[38]

Cardiac entropy! The metaphor to end all metaphors. It will come as no surprise that Rifkin's central point is that society must take into account the second law when planning for the future. We now live in a high-entropy world (waste) and must move to a low-entropy world (frugality). As he so ably puts it in the chapter entitled "Values and Institutions in an Entropic Society,"

> The governing principle of a low-entropy world view is to minimize energy flow. Excessive material wealth is recognized as an irreversible diminuation of the world's precious resources. In the low-entropy society "less is more" becomes not a throwaway phrase but a truth of the highest magnitude. A low-entropy society deemphasizes material consumption. Frugality becomes the watchword. Human needs

are met, but whimsical, self-indulgent desires—the kind pandered to in every shopping center in the country—are not.[39]

Focus on the first sentence. Does he mean entropy or energy? Perhaps both. Well, Clausius said their meanings are nearly identical. So why not entitle the book *Energy*? Yes, a capital idea, for energy is really what worries Rifkin in his chaotic heart. And surely most people would agree that the earth's nonrenewable energy sources are quite finite and will eventually run out. What's more, they are nonrenewable because of the second law. But this position takes a mere two sentences to state, and Rifkin needs to write an entire book. And would most people not agree that to conserve resources is a good thing? Of course. But this position requires only seven words, so Rifkin must enfold it with several hundred pages of second-law unfolding in order to convince his flock that he is *not* talking about the dusty, old-world view of conservation. Old-view strategies such as recycling and nuclear power only delay the inevitable. Who could argue otherwise? But, does not even Rifkin's "frugal," "less-is-more" society only delay the inevitable? Perhaps I am not quick, but do not human beings need to eat? And will this fact alone merely postpone the onset of the energy crisis from one hundred years to a thousand? And is one thousand years infinity? Nay, I say to you it is not.

The prophet is a great advocate of solar power (blinding vision), and with this advocacy comes the admission that the earth is not an isolated system. It receives energy from the sun and is therefore a closed system. If the earth were an isolated system, life would never have arisen and would be impossible to sustain. But once you admit that the earth is not an isolated system, the entire second-law argument begins to unravel. As recounted in previous discussions, the inevitable increase in entropy is only true for isolated systems. One can live with increasing entropy in a closed system as long as energy is imported to sustain the environment. And someday the earth will not merely be a closed system but an open system, receiving minerals from the moon and asteroids.

When this happens the second-law argument will become obsolete.

It is true that the earth is now a closed system and, in terms of material resources, an isolated one. Who would argue against the necessity of conserving these resources except some politicians? Toward the end of his tome, Rifkin writes prophetically, "Like it or not, we are irrevocably headed towards a low-energy society."[40] I have asked the question before: does he mean entropy or energy? Can it be that after 250 pages of unfolding, Rifkin's new world view comes down to "turn off a light bulb today"?

7. ENTROPY AS EPILOGUE

In his review of *Evolution as Entropy*, Lionel Harrison says that the authors have produced something that reminds him of the work of the sixteenth-century alchemist Paracelsus.[41] The analogy is apt, both in regard to Brooks and Wiley and to Rifkin. In one of his astrological works Paracelsus wrote:

> Heaven is man, and man is heaven, and all men together are the one heaven, and heaven is nothing but one man. You must know this to understand why one place is this way and the other that way, why this is new and that is old, and why there are everywhere so many diverse things. But all this cannot be discovered by studying the heavens. . . .
>
> We, men, have a heaven, and it lies in each of us in its entire plentitude, undivided and corresponding to each man's specificity. Thus each human life takes its own course, thus dying, death, and disease are unequally distributed, in each case according to the action of the heavens. For if the same heaven were in all of us, all men would have to be equally sick and equally healthy. But this is not so; the unity of the Great Heaven is split into our diversities by the various moments at which we were born.[42]

This passage contains an undeniable poetic beauty. In its theme of diversity within unity I see a distinct antientropic sentiment. No, I have that backwards. In its theme of unity within diversity I see a clear proentropic philosophy of becoming, not being. No, I still do not have it quite right. In its folly of mistaking a paradox for a discovery, a metaphor for a proof, a torrent of verbiage for a spring of capital truths, and oneself for an oracle, I see Rifkin, I see Brooks and Wiley. And I am willing to wager that after another four hundred years it will be Paracelsus, not they, who is remembered. Not for his science but for his poetry.

ACKNOWLEDGMENT

I am glad finally to have the opportunity to thank my close friend and colleague Dilip Kondepudi for a decade of conversations on entropy and all other matters.

5 · A MEMOIR OF NUCLEAR WINTER

1

On October 30, 1983, a *Parade* magazine article by Carl Sagan introduced Americans to a new form of apocalypse called "nuclear winter." With his article, Sagan had hoped to alert the public to a danger he felt real and relevant to our thinking about nuclear war. Everyone knows how successful he was. The vision of Earth slowly freezing under a twilight sun has found a place in our nuclear nightmares, somewhere beside that of Slim Pickins riding the H-bomb to Armageddon.

This essay is also about nuclear winter, but I will not be concerned so much with the scientific details as with the scientific community. It is also a brief chronicle of some experiences I have had with a group of scientists who dedicate much of their lives to the prevention of nuclear war. In the old chronicles, the monk-storyteller would always begin with a disclaimer that he was "too witless to embellish or ornament the truth" and then go on to describe miraculous and unbelievable events. The immediate result of the disclaimer is that you don't trust a word he says. So I won't make one. Nonetheless, I will try to stick as closely as possible to events that I have witnessed firsthand. You may regard my perspective as one of a scribe who sits at the feet of the participants in the story. If there is something he didn't see, he gets a report from the players. Occasionally he stands up to take a small part.

I shall make my position clear at the outset: as far as anyone can

Bibliographic sources are listed at the end of the essay.

estimate, a full-scale nuclear exchange would kill between 600 million and one billion people outright. For me the necessary calculation has been done; the possible annihilation of one-quarter of humanity and the collapse of civilization are sufficient reasons to prevent nuclear war.

The magnitude of such numbers is not easily imagined. This is, to a great extent, why the nuclear winter campaign has been so successful. The aftermath of a nuclear war now becomes much more vivid. One billion is merely a one followed by a string of zeros. Nuclear winter is an apocalyptic vision. Whether the prophesy proves correct is another question, but it is precisely the sort of apocalyptic vision that has seduced the darker side of the human imagination at least since the Revelation of St. John. To be successful, an apocalyptic vision must be swift, simple, and dramatic. Nuclear winter satisfies all three criteria. Nothing swifter in the way of current warfare can be imagined than a nuclear exchange between the United States and the Soviet Union. The idea is perversely simple. Dust kicked up by the explosions and soot injected into the atmosphere by subsequent fires would virtually extinguish sunlight for months on end. And the end is dramatic. Temperatures would drop to subfreezing levels on a global scale for a time sufficiently long to threaten all remaining life on Earth with extinction.

2

The history of this apocalypse depends on who tells it. In 1982, Paul Crutzen and John Birks published the first estimate of the amount of soot which might be injected into the atmosphere from forest fires ignited during a nuclear exchange. Their paper appeared in a special issue of *Ambio* devoted to the global consequences of nuclear war. *Ambio* is a journal published by the Royal Swedish Academy of Sciences, and the entire special issue has been reprinted by Pantheon Books under a new title, *The Aftermath*. To estimate the amount of soot released by forest fires during a nuclear war is a thankless task. You require many inputs, some of which can only be guessed: the area burned, the fuel available in forests, the amount of smoke emitted per kilogram of fuel burned, and so on. The Crutzen and Birks calculation was very simple—what we call a back-of-the-envelope calculation—but it had the virtue that it directly reflected the uncertainties in their inputs. They concluded that noontime sunlight would be attenuated by a factor of 2 to 150. A light cloud cover attenuates sunlight by more than a factor of two, so at the lower end of Crutzen and Birks's estimate you would not expect a large climatic effect. If the upper end were more likely the result would be disastrous and a nuclear winter would surely result.

Under a nuclear attack you would of course expect cities to burn as well as forests. The data on urban fuel availability were, however, insufficient for Crutzen and Birks to make any estimate whatsoever of city smoke emissions, but they guessed that the amount of city smoke should be comparable to that of forest smoke. If this were true, the likelihood of a nuclear winter would be greatly increased. The central message of Crutzen and Birks's paper is that the potential for climatic change exists, and this conclusion is difficult to avoid.

In October 1983 the Sagan article appeared in *Parade*, and two months later appeared the now-famous *Science* paper, "Nuclear Winter: Global Consequences of Multiple Nuclear Weapons Explosions," by Richard Turco, Brian Toon, Thomas Ackerman, James Pollack, and Carl Sagan. The paper is usually referred to as TTAPS, sometimes, and not without irony, as "et al. and Sagan." The TTAPS calculation coupled a nuclear war scenario to a one-dimensional computer model of the earth's atmosphere. The real atmosphere is three-dimensional and to model it properly requires a three-dimensional simulation. Three-dimensional computer codes are, however, complex and costly, and the TTAPS group began as all scientists would by simplifying the problem as much as possible. They took the atmosphere to be a one-square-centimeter column of air above the earth's surface. This approach ignores many effects, such as wind, but is useful as a first approximation to what might actually occur in the real atmosphere after a nuclear exchange. The computer code allows the modeler to adjust the input parameters, such as the area burned under an attack and the fuel consumption—the same sort of quantities necessary for the Crutzen and Birks calculation. The nuclear war part of the code then generates soot and dust, enabling the atmospheric model to calculate the expected temperature change on the planetary surface. Even in their less severe scenarios, the TTAPS results showed temperature drops to far below freezing. I will talk more about a few of the scenarios as we go along, but for now these details are not crucial. The important message of the TTAPS paper was brought home in a companion article by Paul Ehrlich, John Harte, Mark Harwell, et al., entitled ''The Long-Term Biological Consequences of Nuclear War.'' If the TTAPS conclusions were correct, the aftermath would be a global disaster. Any survivors of the initial holocaust would witness widespread extinction of plant and animal life and their own extinction would not be ruled out.

In late 1983 and early 1984, a number of other studies were published or distributed that evidently confirmed the TTAPS conclusions. The principal among these were a Lawrence Livermore Laboratory report by Mike MacCracken, who performed both one- and two-dimensional simulations; a USSR Academy of Sciences preprint by Vladimir Aleksandrov and Georgi Stenchikov, who presented a summary of their three-dimensional model; and a *Nature* article by Curt Covey, Stephen Schneider, and Starley Thompson of the National Center for Atmospheric Research (NCAR), who also reported the results of a three-dimensional simulation.

The apparent confirmation of the TTAPS calculations has led to a number of public statements, notably by Sagan and Ehrlich, that imply the TTAPS predictions are ''robust'' and that further research will only make them worse. Robust, in the scientific sense, has much the same meaning as it does in other contexts. A result is robust if you can kick it, hammer at it, and drop it from a great height and it will not break. At the moment I am chronicling history so I do not pause to debate the robustness of the TTAPS results. The claims of Sagan and Ehrlich et al., however, raise two questions. The first, and least important, is scientific: Why did the Crutzen and Birks calculation indicate outcomes that varied from mild to severe, while the subsequent calculations all pointed to catastrophe? The second question has a relevance far beyond the realm of nuclear winter: What constitutes scientific confirmation and to what extent may any protagonist construe lack of public rebuttal as assent?

The history of nuclear winter as I have just presented it is similar to versions found elsewhere in the popular literature. But I am less interested in history than in prehistory, that which has gone before. And I am less interested in prehistory than in nonhistory, that which has vanished into lapses of memory and under the weight of words.

Any subject is antedated by prehistory and surrounded by non-history, a fact as true of science as of politics. There are many reasons for this, most of them not malicious, just careless, not sins of commission but sins of omission. It is, after all, a fact of scientific life that results which are not trumpeted are not heard. The Soviet physicist Zel'dovich is fond of saying, "Without publicity there is no prosperity." It is also true that scientists have a child-like outlook on the world. Curiosity and the ability to see a thing in simple terms are facets of childhood; but so is the conceit that history began yesterday, and with it a reluctance to admit that an-other might have had a bright idea before we walked on stage. This human foible is forgivable: a result is always more impressive when it is new and shiny than when it is dimmed by the shadows of forgotten ancestors. In his *Parade* article Sagan writes:

These discoveries, and others like them, were made by chance. And now another consequence—by far the most dire—has been uncovered, again more or less by accident.

The U.S. Mariner 9 spacecraft, the first vehicle to orbit another planet, arrived at Mars in late 1971. The planet was enveloped in a global dust storm. As the fine particles slowly fell out, we were able to measure temperature changes in the atmosphere and on the surface. Soon it became apparent what had happened. . . .

After [the Mariner mission] I and my colleagues, James B.

Pollack and Brian Toon of NASA's Ames Research Center, were eager to apply these insights to the Earth. In a volcanic explosion, dust aerosols are lofted into the high atmosphere. We calculated how much the Earth's global temperature should decline after a major volcanic explosion and found that our results (generally a fraction of a degree) were in good accord with actual measurements. Joining forces with Richard Turco, who has studied the effects of nuclear war for many years, we then began to turn our attention to the climatic effects of nuclear war.

Paul Ehrlich, in his *Amicus* article of 1984, writes:

It has been evident to ecologists for at least a decade that the potential environmental consequences of a large-scale nuclear war were being underestimated or ignored in government-sponsored evaluations. A 1975 National Academy of Sciences study of the *Long-Term Worldwide Effects of Nuclear Weapons Detonations*, for example, failed even to mention many of the ecological problems that would result.

But until very recently there has been little interest in the problem. . . . In an attempt to call attention to risks being ignored in an increasingly unstable arms race [John Holdren, my wife Anne, and I] wrote a section on the ecology of thermonuclear warfare for our 1977 textbook *Ecoscience*, and Anne and I put one in *Extinction* in 1981. But none of the many dozens of reviews of these two books so much as mentioned the topic.

To what extent TTAPS's thinking was started off by the observations of Martian dust storms I am not in a position to say, and the 1975 NAS report was deficient in many respects. On the other hand, in 1963 Tom Stonier published a book, *Nuclear Disaster*, of which several chapters were devoted to the ecological consequences of nuclear war. At about the same time, a Hudson Institute report on the same topic by R. U. Ayres was in preparation, and in 1966 there appeared a Rand study by E. S. Batten dedicated to ''the popular question, 'does the detonation of nuclear weapons affect the weather?' '' Many of the effects currently under investigation are discussed in these works. All try to estimate the amount of dust that would be injected into the stratosphere as a result of a full-scale exchange and the subsequent cooling. The cooling estimates are made by comparing the amount of dust injected with that of a major volcanic eruption. Similar estimates for individual atomic tests date back at least to 1955. Whether the "volcano analogy" is relevant to nuclear winter has been debated, but it is naturally the first thought that pops into any scientist's head. Indeed, the first suggestion that volcanos might cool the Earth's atmosphere was apparently made by Benjamin Franklin. Credit for the first calculation that predicted from basic principles the temperature drop due to a volcanic eruption is usually given to W. J. Humphreys, who derived his result in 1913. It can be found in his classic text *Physics of the Air*, which is still in print.

In his report, Batten also discusses snow and ice feedback, recently reintroduced to the nuclear winter debate. Stonier speculates on the triggering of a new ice age, though he doubts the possibility. Batten mentions that soot from forest fires could cause substantial reductions in temperature and points to the Alberta smoke pall of 1950, which he claims lowered the temperature in Washington, D.C., by 10°F. If he had taken a further step and estimated the quantities of smoke generated by a nuclear war, he would have predated the Crutzen and Birks calculation by fifteen years. Other ecological consequences of nuclear war are also discussed in these works: soil erosion due to forest fires, epidemics, plague.

The tone of the authors varies and, if the stereotypical rules hold, I suspect their political biases vary as well. Batten sticks close to the problem and simply states what is known and what is not. He concludes that "large climatic impact is possible but its magnitude is unknown." Ayres tends to minimize the consequences and even scoffs at doomsayers. Twenty years later his voice finds resonance in that of Edward Teller. In order to maximize public impact, Stonier's book contained a preface by Gerard Piel, publisher of *Scientific American*. Nevertheless, *Nuclear Dis-*

aster sank into oblivion. I am sure he would sympathize with Ehrlich's feeling of neglect.

Ehrlich's proprietary tone is echoed in TTAPS's 1984 *Scientific American* article: "Although in the wake of such a war the social and economic structure of the combatant nations would presumably collapse, it has been argued that most of the noncombatant nations—and hence the majority of the human population—would not be endangered, either directly or indirectly."

The TTAPS argument would be more convincing if they offered an example of a strategist who held both that the social and economic structure of the combatant nations would collapse and that the noncombatant nations would remain unharmed. I cannot remember ever having seen such a statement. On the other hand, counterexamples are plentiful. Henry Kissinger in 1958: "Our current military policy is based on the doctrine of massive retaliation—that we threaten an all-out attack on the Soviet Union in case the Soviet Union engages in aggression anywhere. This means that we base our policy on a threat that will involve the destruction of all mankind." An official statement of the Soviet government in 1963: "Thermonuclear war will have catastrophic consequences for all peoples, for the whole world. All countries, even those that survive the war would be set back in their development by decades, if not centuries." Herbert York and Jerome Wiesner in 1964: "It is very difficult to make precise estimates, but it seems that a full nuclear exchange between the U.S. and the U.S.S.R. would result in the order of 10,000,000 casualties from cancer and leukemia in countries situated well away from the two main protagonists. In addition, genetic problems that are even more difficult to calculate would affect many, many millions of others—not only in this generation, but for centuries to come." The *Ambio* issue in which Crutzen and Birks's article appeared is entirely devoted to the global consequences of nuclear war. One paper by Yves Laulan is entitled "Economic Consequences: Back to the Dark Ages." Laulan, and H. W. Hjort in his companion piece, "The Impact On Global Food Supplies," make a fairly obvious point that cannot be stressed too much: North America and Europe export most of the world's food supplies, not to mention technology. If North American and European civilization collapses, it is folly to think the rest of the world will remain unaffected. The number of Third World starvations alone that would result is probably uncalculable.

I suspect that if any strategist ever held the schizophrenic position TTAPS claims for hm, it was because he refused to face the obvious. For this reason the obvious should be proclaimed loudly and clearly. But as presented, the TTAPS argument carries with it the uneasy ring of a straw man.

Even if Stonier's voice had been heeded twenty years ago, or if the Rand and Hudson Institute reports had found their way beyond the labyrinths of Washington and libraries, it is perhaps too much to expect that they would have had an impact on the arms race; very little does. In any case, Stonier's voice remained in the wilderness and the other reports on archival shelves. Sagan, Ehrlich, and their colleagues must therefore be given great credit for bringing the potential apocalypse of nuclear winter before the public's attention. But few ideas spring wholly formed from the head of Zeus. That the early studies did not make detailed numerical calculations is not so important. Calculations are plentiful while ideas are few. TTAPS reference a few of the precursors to nuclear winter in their unpublished drafts. They would not be due any less credit if they mentioned some of these lost works in public, or even hinted at their existence.

3

I am now going to leave the catacombs of prehistory and nonhistory and begin a chronicle of more recent events. Since these events are closer to personal experience, I will simply tell the story as I saw it.

My own nuclear winter education emerged from a fusion of other interests. By training I am a physicist, specifically a cosmologist. With many other physicists I share a primordial sense of guilt for the nuclear dilemma mankind faces and, like any other human being, as I grow older I experience an increasing awareness of the world beyond the horizons of my narrow specialty. For a number of years I have held a particular love for the Russian language and culture. The land of Pushkin, snow, and white nights has drawn me back repeatedly and, to date, I have spent over a year in the Soviet Union.

When living in the Soviet Union, one experiences firsthand the obsessive secrecy, the lack of information about events both within and without the country and, with most relevance to nuclear arms and winter, the reliance on Western technology. I could only commiserate with my colleagues at the Shternberg Astronomical Institute who, in an unheated building, retrieved telescope data on paper tape, who were allowed to submit their punched-card Fortran codes two days a week and collect output one day a week. The situation was brightened somewhat by the arrival of a microcomputer at the university, but a guard was stationed next to it.

If such hardships have a surreal air to them, so do many of the statements the American government makes about the Soviet Union. U.S. strategists maintain that the Soviet Union has elaborate civil defense plans which would be activated in the event of a nuclear attack and ensure the survival of the Soviet population. Indeed, it is impressive—even miraculous—that U.S. experts can calculate how many hours the Soviets would require to evacuate Moscow. I have never met a Russian who knows of any civil defense plans.

Communications gaps are more real than missile gaps and I returned from my last stay in Moscow wanting to use my education and experience for the cause of peace. The Russians have a saying, "If you call yourself a mushroom then jump into the basket." At home a friend suggested that I contact the arms-control group at Princeton University's Center for Energy and Environmental Studies. I took his advice.

CEES, as the Center is called, is housed in the von Neumann building at the far edge of campus. Fifteen years ago, the same building was occupied by the Institute for Defense Analysis. Protests following the bombing of Cambodia led to IDA's removal from the university and its eventual replacement by CEES. This irony is not lost on CEES staff, who form a multidisciplinary and admittedly liberal corps devoted to solar energy research, hazardous waste disposal, energy-efficient architecture, and arms control.

The arms-control group, like the Center in general, is a haven for physicists who have developed a social conscience. Frank von Hippel is an ex-particle physicist who until 1984 served as the elected chairman of the Federation of American Scientists, a lobbying organization devoted to nuclear disarmament. Barbara Levi, a research staff member, is also an ex-particle physicist, and Hal Feiveson, while now a political scientist, has a physics background as well. Historian Richard Ullman of Princeton's Woodrow Wilson School for International Affairs is their close collaborator. Although they will undoubtedly find it highly presumptuous of me, I will refer to this band as the "World Savers."

The World Savers consider their job to be twofold. First, they are actively engaged in understanding the technical aspects of nuclear policy and arms control. Reactor safety studies, the global control of plutonium, and the effects of limited nuclear war are among the subjects dealt with by the World Savers. At present, they are investigating the feasibility of a "finite deterrence" defense strategy. "Finite" is meant to stand in opposition to "infinite," the current size of our strategic arsenals, and the program asks whether it is possible to deter an enemy attack using far fewer missiles than are currently deployed. The second part of the World Savers' job is based on an acknowledgment of a responsibility to communicate their results to policymakers and to the general public. The World Savers frequently give public lectures, contribute technical and popular articles to a number of magazines, and testify before Congress.

My introduction to the concerns of the World Savers was largely through their weekly lunch seminars. Topics that range from anti-submarine warfare to the military strategy of World War I to the problem of succession in the Soviet leadership are discussed over hoagies. In addition to CEES World Savers, other World Savers are frequent participants: Nobel laureate Philip Anderson from the physics department, Freeman Dyson of the Institute for Advanced Study, and a host of political scientists, activists, and military commanders. The seminars are always lively, sometimes contentious, and often continue for hours.

Shortly after I began to visit the Center on an informal basis, the nuclear winter story broke in *Parade* magazine. The following day the Conference on the Long-Term Biological Consequences of Nuclear War opened in Washington, D.C., where Sagan and Ehrlich officially announced the results of the TTAPS study. While Frank von Hippel in particular had known something was up, it is fair to say that even the World Savers were surprised at the magnitude of the publicity and that the results were first published in *Parade*. Other than forgotten precursors and the Crutzen and Birks article, no scientific paper on the subject had yet appeared. This put scientists in an awkward position. Within days, friends and layman began to ask for our opinion on the subject and we could not give an informed response.

Later that same week a preprint of the now-famous TTAPS *Science* paper arrived from Washington. Several aspects of this preprint puzzled us. TTAPS had taken as its "baseline" a 5,000-megaton exchange of which 1,000 megatons were targeted against cities. The baseline case changed the average temperature of the northern hemisphere from roughly 15°C to − 20°C, a 35° drop. It is this scenario which has given rise to the term "nuclear winter" and the vision of ruined cities that lie beneath crimson skies and swirling snow. But at the same time TTAPS presented a "city attack," in which a total of 100 megatons directed against cities gave essentially the same result as the baseline case. It is this scenario which has often been termed the "threshold" scenario and has given rise to speculation that even 100 megatons, or a limited nuclear war, could trigger nuclear winter. We asked ourselves how a 100-megaton exchange could produce virtually the same effect as a 5,000-megaton exchange and could not find an obvious answer; in the TTAPS preprint there is no explanation. Generally, the *Science* paper is a summary of results. Any explanations were to be found in reference 15, the longer version of their paper, which I had not been shown at the time. Reference 15 remains unpublished.

During roughly the same period, Barbara Levi also received a copy of the Soviet nuclear winter study headed by Aleksandrov. Although the paper is not readily available, Aleksandrov presents the main results in the recently published book *The Cold and the Dark*, which constitutes the proceedings of the Washington conference. Certain aspects of the Soviet investigation were unclear as well. In particular, their results showed that, after a year, the atmosphere would heat up beyond the original temperature almost as much as it cooled down. Figure 8 of their paper, which shows this overshoot, is not presented in *The Cold and the Dark*. After my year in Russia I was also interested to see that not only was the Soviet model based on a 1971 American computer code, but that every reference in the paper, with the exception of one to their own work, was to an American source.

The only way out of an information vacuum is to pretend you are an expert or become one. The World Savers are not politicians and thus chose the second option. To begin the education process, Jerry Mahlman was asked to give a seminar on nuclear winter. I was invited to sit in. Mahlman is an atmospheric physicist who has recently been appointed director of the National Oceanic and Atmospheric Administration's Geophysical Fluid Dynamics Laboratory, located at Princeton University's Forrestal campus. He had also participated in a private conference on nuclear winter held in Cambridge, Massachusetts, the previous April and was on the review board of the National Academy of Sciences' nuclear winter study which was then in the works. Mahlman stressed that he held great respect for the abilities of Turco and his colleagues— "these guys aren't amateurs"—but severely criticized a number of the assumptions that went into the TTAPS model. He also introduced us to the work of Mike MacCracken at Lawrence Livermore Laboratory who, as I have said, was engaged in similar calculations. By the end of the seminar we were not experts, but it was apparent that the TTAPS results were preliminary and much room was open to scientific criticism.

We were further educated in early December 1983, when a Soviet arms-control delegation arrived in Princeton to meet American counterparts. Among them was Aleksandrov. In *The Cold and the Dark* Aleksandrov claims that the Soviet results are "the same" as the NCAR conclusions and, by extension, those of TTAPS. In conversations with scientists he is much more frank about the limitations of their computer model. He acknowledged from the start that the temperature overshoot I mentioned is an artifact of their model, not of the atmosphere. This fact is not reported in *The Cold and the Dark* either. I don't want to make too much of the overshoot; I simply feel that mistakes should be reported along with successes.

Also visiting with the Soviet delegation was Sergei Petrovich Kapitza, a trained physicist, editor of the Soviet edition of *Scientific American, V Mire Nauk*, and host of a popular-science show on Soviet television. A few days after the delegation's departure, I received a prepared statement by Kapitza that he was to have read in Washington. He is an eloquent writer in English and it must be an undiluted pleasure to read him in Russian. His thoughts are not entirely hs own—the Martian dust storms are there, the claims that the Soviet results confirm the American results. There are also the ritual obeisances to scientific socialism and to the Soviet government, but his essay nonetheless leaves an impression of great sincerity. And that is for one simple reason: he leaves himself out of it.

4

In late spring, 1984, Frank von Hippel asked Barbara Levi and me to spend the summer reviewing the nuclear winter calculations in order to educate the World Savers and to arrive at whatever conclusions we could. Until this time I had been merely a frequent guest at the Center. Now I became a temporary World Saver.

In making his request, von Hippel expressed a concern that many scientists have had about the nuclear winter publicity from the beginning: If nuclear winter should turn out to be nonexistent, or even less severe than claimed, the effort to mount a disarmament campaign on it could backfire badly. The credibility of scientists would be damaged, as would the disarmament movement.

For this reason, many liberal scientists feel the safer course is to check the nuclear winter calculations themselves, rather than leave the job to the Pentagon and to the extreme right. The reader may ask why political leanings should bias scientific objectivity. It is a very good question.

The Center did not have a large computer model at its disposal so our job was to some extent secondary. When a field expands as rapidly as nuclear winter, new papers appear almost weekly and, in this case, most were unpublished. Nevertheless, I think Levi and I managed to read almost every word written on the subject; we checked calculations and did our own such as we were able; we spoke to many specialists around the country and invited several to give talks at the Center.

It quickly became apparent that a scientific consensus on nuclear winter does not exist. Unless it is a consensus not to have a consensus. I have rarely seen such an intense debate over a scientific topic. There was no point not open to dispute.

One question, however, was answered by the published version of the TTAPS *Science* paper which appeared at the end of 1983: How did a 100-megaton city attack produce essentially the same results as the 5,000-megaton baseline attack in which 1,000 megatons were targeted against cities? This city or threshold scenario has received much public discussion and is worth a few words. In *The Cold and the Dark* Sagan writes:

> Perhaps the most striking and unexpected consequence of our study is that even a comparatively small nuclear war can have devastating climatic effects, provided cities are targeted (see case 14 in figure 2; here the centers of 100 major NATO and Warsaw Pact cities are burning). There is an indication of a very approximate threshold at which severe climatic consequences are triggered—by 100 or more nuclear explosions over cities [in terms of smoke generation].

There is no further explanation in the text or in the figure captions. Nonetheless, both Kapitza and Ehrlich repeat Sagan's statement almost verbatim in their essays, and speculation on the existence of a nuclear winter threshold has found its way to television and the *New York Times*. Unfortunately, Sagan's account of the TTAPS results is both misleading and incorrect. A close reading of their paper shows that, for the threshold attack, the 100 megatons are assumed to be distributed over one thousand city centers, not one hundred. Because city centers contain more potential fuel than suburbs or forests, such a distribution targets only the highest fuel sources. Sagan also does not mention that the amount of smoke emitted from each city-scenario target was taken to be about five times the value used in the baseline case. The selective targeting plus the fivefold increase in smoke emissions brings the total threshold smoke up to 60% of the baseline smoke. That the threshold and baseline scenarios produce similar climatic effects is not surprising; the only surprise is the term "threshold" to describe the situation. A correct version of Sagan's statement might read: ". . . a comparatively small megatonnage can have devastating climatic consequences provided 1,000 city centers are targeted and smoke emissions are five times higher than our baseline values." I have replaced the words "small nuclear war" by "small megatonnage" because an attack on 1,000 city centers targets essentially every city in the United States, the Soviet Union, and Eastern and Western Europe with populations over 100,000. This is not a limited nuclear war.

For the purposes of this essay I need not go into further specifics of the TTAPS scenarios. The results of our survey indicated that, within the limits of current knowledge, almost any answer is possible. The nuclear winter calculations were not robust. This is not to say that nuclear winter is ruled out, nor does it mean the TTAPS calculations are wrong. It means simply that I could pick assumptions and data which are as valid as the TTAPS choices, and nuclear winter would essentially vanish. Reports from the nuclear winter conference held in Erice, Sicily, support this contention. After we examined as much data as we could find, we concluded that the original Crutzen and Birks calculation gave as accurate a prediction as any made thereafter.

I have chosen the word "accurate" deliberately. The TTAPS and subsequent calculations are much more precise than the Crutzen and Birks calculation but this is not the same thing. In my undergraduate physics laboratory, a cartoon hung on the wall that showed a disgruntled archer facing a target where five of his arrows have all landed in a very tight cluster on the outer ring—far from the bull's eye. The caption read "Precision is not accuracy." A computer can make an answer more precise; it cannot make it more accurate. And Batten's conclusion of 1966 remains true today: "Large climatic impact is possible but its magnitude unknown."

It is just this uncertainty that should alarm us; to the extent of our knowledge almost anything could happen in the wake of a nuclear war. A result that is not robust may certainly turn out to be wrong. But it could also turn out to be right, or nearly so. For this reason I do not want to sound optimistic. The 1815 eruption of Tambora produced an average temperature drop around the globe of less than one degree. This is far from a nuclear winter but the results were disastrous. Cold snaps in June of the following year led to crop failures and widespread famine. It is not inappropriate that 1816 is remembered as "The Year Without a Summer." A technical debate over whether the final answer will be 5° or 35° is important if you are worrying about disaster versus extinction. It is important for scientists who are trying to understand whether one study confirms another. It is also important for those who worry about the political side effects of exaggeration. But it is not so important if you believe that even a one-degree temperature drop is unacceptable.

So, the response to the nuclear winter calculations depends on the question you are asking. I have claimed the calculations are not robust; others claim they are. As a scientist seeking truth and worried about credibility gaps, I am intensely interested in the resolution of this dispute; as a person opposed to nuclear war on first principles, I find the resolution somewhat more academic. But let me continue for a moment as a scientist and ask again the question I asked before: To what extent is it meaningful to say the results of one calculation confirm those of another? Though I pose the question as a scientist I would like it to have broader applicability than to nuclear winter alone. In setting it, I am not really interested in whether the TTAPS results are robust or whether they are not.

I am interested to discover on what basis a claim for robustness can be made.

Sagan believes the TTAPS calculations have been confirmed. In his *Foreign Affairs* article, which appeared early in 1984, he writes: "The new results have been subjected to detailed scrutiny, and half a dozen confirmatory calculations have been made. A special panel appointed by the National Academy of Sciences has come to similar conclusions." Every paper that appears in an established scientific journal has passed through a peer review in which one or more fellow scientists have decided that the work merits publication. This does not mean that every scientific paper is right. To the contrary, all scientific papers must in some sense be wrong, if for no other reason than every paper is based on assumptions that will break down under certain conditions. A referee's job is to filter out crackpots, check for obvious errors, and suggest improvements in the exposition. The scrutiny to which Sagan refers includes not only a standard review of the *Science* paper before publication but a closed conference held in April 1983, six months before the results were announced in *Parade*. The expressed purpose of this conference was to examine the TTAPS calculations. First, let me take up the confirmatory studies. These include the NCAR work by Covey, Thompson, and Schneider; the Aleksandrov and Stenchikov model; and MacCracken's Livermore model.*

The TTAPS baseline case produced a temperature drop of 35° and the other studies predicted about 20°, 20°, and 8°, respectively. In quantum electrodynamics a factor of four would not be regarded as confirmation; it would be regarded as a dispute. Climatology is a much less exact science and it is to some extent a matter of taste whether one regards 8° as a confirmation of 35°. To be fair, I should mention that TTAPS estimated that the effects of oceans might reduce their calculated temperature drop by 30% to 70%, which would put the answer in the same ballpark as the others. By the same token I should add that the NCAR results

---

* I do not know to what other confirmatory studies Sagan is referring in his *Foreign Affairs* statement.

showed the 20° change only if the nuclear exchange occurred in summer. If the presumed exchange took place in winter, the NCAR simulation strongly suggests a very small effect of at most a few degrees. This also appears not to be reported in *The Cold and the Dark*.

But in this discussion the essential point lies elsewhere. Some problems are such that no matter what inputs you choose, the output remains unchanged. We have all had magicians ask us to pick a number at random and then perform elaborate arithmetic operations on it. Not knowing the original number, the magician miraculously tells us the final answer. These parlor tricks work because certain algorithms always converge to the same result. The nuclear winter calculations are not of this type. The final answer depends very critically on the input values, for instance on the amount of smoke assumed to be injected into the atmosphere. If the quantity of smoke turns out to be ten times less than TTAPS used, nuclear winter will essentially disappear. All the confirmatory studies took as an input the amount of smoke given by the TTAPS baseline case. The results are certainly robust in the sense that, given this amount of smoke and given that it remains in the atmosphere for the amount of time TTAPS claim it does, something drastic will happen to the earth's climate. The major uncertainties—i.e., is this amount of smoke likely and will it stay in the air for months—have simply not been addressed by the confirmatory studies. The recent calculations all converged to the upper end of Crutzen and Birks's estimate first because they added the effects of cities and, second, because they took smoke emission to be in the higher regime of allowed values. Precision is not accuracy.

Memory is not always accurate. Nevertheless, I believe Einstein once said, "You should make things as simple as possible but not too simple." The Earth's atmosphere is extremely complex and the nuclear winter models are very simple. We can be sure that whatever happens will be much more complicated than what has so far been predicted. In at least one nuclear winter model Einstein's dictum has been violated, if unintentionally. I have already mentioned that the temperature overshoot in the Soviet model was due to severe computer constraints. In addition,

Aleksandrov and Stenchikov gave dust the optical properties of soot. Because soot absorbs light much more effectively than dust, this error overestimates the reduction of sunlight reaching the Earth's surface by a large factor. In a recent issue of *Science*, the Soviet model received some harsh, if accurate, words from Richard Turco and Starley Thompson. Both go on to say that the Soviets have contributed little to the international nuclear winter program. I submit that the disappointment expressed by Turco and Thompson is a result of unrealistic expectations and would encourage them to continue collaboration with the Soviets if for no other reason than to avoid the frequent misunderstandings and surprises that dog U.S.-Soviet relations. It is better to know what they are doing than not to know what they are doing. In the meantime, it is scientifically unwise to regard the Aleksandrov and Stenchikov results as strong confirmation of any other study, and I find it ironic that Sagan continues to do so while his principal coauthor does not.

Degree of confirmation is, as I have indicated, largely a degree of taste. One could argue that the Soviet study does weakly lend support to the TTAPS conclusions. I am willing to accept this point of view even if I do not subscribe to it myself. Each of us draws the line where our instincts prohibit further retreat. In my own case, the line is drawn before citation of the National Academy of Sciences/National Research Council Study.

At the time Sagan's *Foreign Affairs* article appeared, the NAS/NRC preliminary draft had not been passed by the review committee.* It was confidential and not to be quoted or cited. In any case, the two principal authors of the TTAPS paper were members of the NAS drafting team, and the NAS report relies so heavily on the TTAPS results that it is difficult for me to regard it as an independent investigation.

As it happens, the final version of the NAS/NRC study has recently been released and Sagan continues to cite it for support. The NAS committee clearly states

---

* A committee member has recently remarked to me that it had not yet been written, though I do not know the exact timetable.

that unless one or more of the effects lie near the less severe end of their uncertainty ranges, or unless some mitigating effect has been overlooked, there is a clear possibility that great portions of the land area of the northern temperate zone (and, perhaps, a larger segment of the planet) could be severely effected. Possible impacts include major temperature reductions (particularly for an exchange that occurs in the summer) lasting for weeks, with subnormal temperatures persisting for months.

The report also states that "the committee cannot subscribe with confidence to any specific quantitative conclusions drawn from calculations based on current scientific knowledge." And it emphasizes that "A more definitive scientific statement can be made only when many of the uncertainties have been narrowed, when the smaller scale phenomena are better understood, and when atmospheric response models have been constructed and have acquired credibility for the parameter ranges of this phenomenology." I will be forgiven if I also quote a slightly earlier passage in the study: "Some reviewers of earlier drafts of this report . . . suggested that at present the only scientifically valid conclusion would be that it is not at this time possible to calculate the atmospheric effects of nuclear war."

5

With one or two exceptions, the above remarks are scientific in the broadest sense of the term. They stand or fall on their own merit, independently of who made them. Unfortunately, the nuclear winter debate has not avoided taking on an *ex cathedra* and *ad hominem* character and to this extent it has become unscientific. In a recent rebuttal to *Nature* editor John Maddox, TTAPS wrote:

> Our findings on what we have called nuclear winter evolved from, and were partly calibrated by, 12 years of related research on Martian dust storms, the climatic consequences of volcanic explosions on Earth and the possible col-

lision of an asteroid or cometary nucleus with the Earth at the time of the Cretaceous/Tertiary extinctions.

Given the importance and sensitivity of the subject, we took extraordinary measures to have our calculations reviewed by a large number of experts in atmospheric physics and chemistry, at a meeting specially convened for this purpose in April 1983, and by other means—well before the submission of the paper for publication. Our article refers to 95 published scientific papers and reports in which further details can be found. . . .

We hold . . . that open and informed debate is the only responsible approach, given the gravity of the potential climatic catastrophe we believe we have uncovered. . . .

. . . By his statements, Maddox also seems to be unaware that climatic effects of volcanic explosions are caused principally by sulphuric acid aerosols, not by silicate dust.

Any layman or scientist who is not familiar with the details of nuclear winter and who reads the above passage attentively will be perplexed and will want to question three points of logic. First, how can a technique be calibrated against an event—the collision of an asteroid with the Earth 65 million years ago—that may not have happened? Second, if the primary effects of volcanic explosions are caused by sulphuric acid aerosols and the nuclear winter effects by soot and dust, then of what relevance is a calibration between volcanoes and nuclear winter? These two questions attempt to drag us back to the scientific domain, but the third offers a transition into a new realm: of what significance is the fact that twelve years of research were involved and ninety-five references cited? Basically none. In my own field, the citation of ninety-five references would undoubtedly mean that most of them haven't been read. But I am willing to give TTAPS the benefit of the doubt here. As to twelve years of research—well, twelve divided by five isn't so much where science is concerned (and I refuse to take seriously the notion that TTAPS are claiming sixty man-years). A small research project typically takes a dozen man-years and a large-scale program may take hundreds and thousands. I can say from personal experience, as can any scientist, that interminability does not provide an exemption from failure.

I have now mentioned the peer review conference of April 1983 several times. It was held in Cambridge, Massachusetts, and is described by Ehrlich in his *Amicus* article. I will let him tell the story as he saw it:

> The notion of setting up an important conference in less than three months was appalling but we all thought the seriousness of the issue made it imperative. Everyone agreed, however, that these meetings should be closed to press and public. Great harm could result if conclusions were reached before they were reviewed thoroughly or if speculations were mistranslated into fact. [Peter] Raven, one of the busiest biologists in the nation and one of the leaders in the battle to save tropical rain forests, was prevailed upon to accept the task of organizing the meeting. The rest of us promised to help.
>
> Others were like-minded. About seventy of the nation's top scientists were concerned enough to drop everything and converge on Cambridge in April. For two days, the TTAPS results were given intensive scrutiny by some of the toughest possible critics including Schneider, who specializes in climate modeling; Holdren, whose detailed knowledge of arsenals and arms control qualifies him to evaluate the TTAPS war scenarios; Crutzen and Birks; George Carrier from Harvard, who was then just about to chair a new U.S. National Academy of Sciences Committee on the atmospheric effects of nuclear weapons explosions (the Academy had recognized the inadequacy of its earlier study); Vladimir Aleksandrov of the Laboratory of Climate Modeling of the Soviet Union's Academy of Sciences; Jerry Mahlman, an outstanding atmospheric dynamicist from Princeton; and Robert Cess, an expert in radiative transfer from the State University of New York at Stony Brook.

Ehrlich writes in summation:

> TTAPS's basic conclusions went unchallanged, even though everyone present had hoped they would prove too pessimistic. *It appeared that even the explosion of a 100-kiloton warhead over each of a thousand cites (a 100-megaton war) could bring on a nuclear winter—and that would mean the detonation of only about one percent of the combined present American-Soviet arsenals (some 10,000 megatons).* [Ehrlich's italics.]

Ehrlich has correctly stated the number of explosions assumed in the TTAPS city scenario but otherwise his passage is highly misleading. The enthusiasm he marshals was not universally shared.

Robert Cess visited us in Princeton to give a seminar and has provided us with much useful insight into the nuclear winter problem. I wish to thank him now. He tells us simply that he was invited to the Cambridge conference on the basis of having changed a number in a computer code, gave a talk, and left. He participated in no peer review process. Cess did not challenge the TTAPS findings because he was essentially not there. Ehrlich has presumed his silence betokened consent. Since the conference, Cess has written a paper that shows that some of the TTAPS simplifications must result in an overestimate of the cooling. He calls the Ehrlich article "a snow job."

During our work, Jerry Mahlman also gave us frequent advice and participated in our seminars. Of Ehrlich's summary remark he says:

> It is completely *untrue* that the basic conclusions went unchallenged. Points that I can recall were challenged include: Their smoke altitude to 15 kilometers (Carrier and Mahlman) later reduced; one-dimensionality and neglected effects of "spottiness"; heat capacity of the oceans; assumed smoke optical properties; greatly accelerated smoke transport to the Southern Hemisphere; gloom over the tropics; neglect of condensation cap at the top of smoke plumes.

Do not be waylaid by the technical jargon. The general tenor is evident. It is fair to say Mahlman finds many of the TTAPS assumptions unjustified and some untenable. It is fair to say he is critical of the entire one-dimensional approach. It is not fair to mute his criticisms and thereby imply he agrees with the TTAPS conclusions. He simply doesn't. "The TTAPS group has a high-side bias," he has said, "and I'd be willing to put money on the table that the final answer is down from their result." Mahlman has also accurately expressed the feeling of many scientists regarding the progress of the nuclear winter debate. "In a rational scientific environment, no one would be bothered by the lack of consensus. This is a five- or ten-year research program. The difficulty is that we are being forced to converge to a result in six or eight months."

For perhaps similar reasons, Frank von Hippel, who was on the steering committee of the April conference, resigned in protest. "Sagan was not going to allow discussion of other work and I saw that it would be a sideshow."

George Carrier is reported in a recent issue of *Nature* to have said that he believed both Sagan and Edward Teller have taken the results of the recent calculations "too literally," and that the current models can at best give "indications" of the climatic change following a nuclear war.

Near the end of the summer, we asked Richard Turco to visit CEES to report on the latest nuclear winter research. When two scientists meet for the first time an important, if unspoken, impression is exchanged—whether personalities would allow for future collaboration. My feeling was that Turco and I could get along quite well. He was very receptive to our criticisms and suggestions. Both he and Tom Ackerman have been equally forthcoming with the details of the TTAPS computer code. The *New York Times* quotes Turco as saying, "Initially there was lots of skepticism. People tried to punch holes in it, but that didn't work." Perhaps he feels that way, but with us he displayed such an appreciation of the uncertainties involved that it was difficult to argue with him. He agrees that two days at a conference are not enough to "get down to the nitty-gritty." He freely acknowledges that

Jerry Mahlman "had trouble with the study." Turco and I will continue to have our differences. But I have no quarrel with the remark he made to me late in the afternoon: "The potential for ecological disaster is large."

6

Ehrlich's article, in an obvious sense, has nothing to do with nuclear winter. I have alluded to the famous exchage at the end of Robert Bolt's play *A Man for All Seasons*. Thomas More has refused to give his consent to the king's divorce and is brought to trial for high treason. Cromwell, the prosecutor, argues that the silence of a corpse betokens nothing: "This is silence pure and simple." The silence of a tacit witness to a crime betokens complicity. More's silence betokens "not silence at all, but most eloquent denial."

> MORE: The maximum of the law is: silence gives consent. If, therefore, you wish to construe what my silence "betokened," you must construe that I consented, not that I denied.
> CROMWELL: Is that what the world in fact construes from it? Do you pretend that is what you *wish* the world to construe from it?
> MORE: The world must construe according to its wits. This court must construe according to the law.

This dialogue applies directly to the nuclear winter debate. The majority of scientists have remained silent or spoken with muted voices. Ehrlich has construed "according to the law" that their silence betokens consent. But Cromwell's question is relevant: what do they wish the world to construe?

My answer is that the silence of the majority betokens the silence of a corpse. Most scientists have simply not looked closely at the nuclear winter calculations and most scientists, by virtue of experience, do not express opinions on theories about which they know nothing. This is silence pure and simple. There is a smaller class of scientists, on the other hand, who have thought about nuclear winter and continue to say nothing. These scientists may be likened to Thomas More who, after his conviction, admits that he does oppose the king.

Neither the silent majority nor the Thomas Mores want their opinions and expertise unwittingly co-opted for any cause. In this regard scientists are like anyone else. But unless they break their silence, their opinion will be construed. According to the law.

Ninety times out of one hundred, Freeman Dyson's first reaction to a statement is, "I disagree." It is a critical response but a creative one, for it allows other possibilities to be explored. At this moment I see the purse of the lips and the shake of the head and hear, "I disagree." But this is all right. Over the last few years, Dyson has taught me much and now I am going to use his own lessons against him; it is the best compliment I can pay him. It happens that he has recently written a lecture entitled *Star Wars, Austrianization, and Nuclear Winter.* In his own words:

> When Carl Sagan and his colleagues began two years ago to bring the possibilities of nuclear winter dramatically to the attention of the public, they put professional scientists like me into an awkward position. On the one hand, the professional duty of a scientist confronted with a new and exciting theory is to try to prove it wrong. That is the way science works. That is the way science stays honest. . . .
>
> On the other hand, nuclear winter is not just a theory. It is also a political issue with profound moral implications. If people believe that our weapons endanger not only our own existence and the existence of our enemies but also the existence of human societies all over the planet, this belief will have practical consequences. It will lend powerful support to those voices in all countries who oppose nuclear weapons deployments. It will increase the influence of those who consider nuclear weapons to be an abomination and demand radical changes in present policies. So my instincts as a scientist come into sharp conflict with my instincts as a human being. As a scientist I want to rip the theory apart but as a human

being I want to believe it. This is one of the rare instances of a genuine conflict between the demands of science and the demands of humanity. As a scientist I judge the nuclear winter theory to be a sloppy piece of work, full of gaps and unjustified assumptions. As a human being I hope fervently that it is right. It is a real and uncomfortable dilemma.

Dyson goes on to say that there are three responses to the dilemma. The first is to say that we are scientists second and human beings first, so we will forget our scientific misgivings and "jump on to the nuclear winter bandwagon." The second response, as scientists dedicated to truth, is to criticize nuclear winter as harshly as we would "any other half-baked theory." The third is to realize it won't do any good in the long run to believe a wrong theory but it will not do any good in the short run to attack it publicly, so let us remain silent until the facts become clear. Dyson has chosen the third response. "It is an unheroic and uncomfortable compromise, but I prefer it to either of the simple alternatives. The dilemma is similar to the dilemmas which occur frequently in personal and family life, when the demands of honesty and friendship pull in opposite directions. It is good to be honest but it is often better to remain silent."

Every scientist who has looked more than casually into nuclear winter shares Dyson's dilemma and feels the same awkwardness. Nonetheless, I disagree with him on two counts and therefore arrive at a different response.

The first count is one of logic. Dyson reads a public lecture at the University of Chicago in which he says, "As a scientist I judge the nuclear winter theory to be a sloppy piece of work, full of gaps and unjustified assumptions." A few sentences later he says, "Let us remain silent until the facts become clear." When I read documents from the Defense Technical Information Center, it is always a mental jolt to stumble across, printed in large, bold letters.

BLANK PAGE

This is a self-referential logical contradiction. Dyson's essay produces much the same effect. The source of the contradiction is not hard to find. It lies precisely in the conflict of interests that Dyson himself discusses. Both his wish to remain silent and his urge to speak out have surfaced in the same lecture. I sympathize with his dilemma more than he probably knows but, as the Russians say, "You can't eat the same pie twice."

My second disagreement with Dyson lies in his interpretation that "nuclear winter is a political issue with profound moral implications." That nuclear winter is a political issue cannot be denied. The question is, does it deserve to be one? It is not clear to me at what point a scientific theory should become the basis of a political issue, but my instincts tell me that the science should be established first, the politics second.

It is equally dangerous to base the moral argument on a tentative scientific result. Surely, to kill one billion innocent people is an act so close to absolute immorality that it should be sufficient to guide our moral debates about nuclear war. The proponents of nuclear winter who conjecture four billion dead and conclude a fourfold increase in immorality strike me as akin to the medieval theologians who counted angels on pins and the perfections of God.

Such philosophical quandaries would be academic if they did not lead to distinct responses. I can illustrate how this happens best by resorting to a personal experience. This story will also help explain why my response to nuclear winter is different from Dyson's.

Not long ago, I had the opportunity to give a number of talks on nuclear winter in Austin, Texas, including a luncheon seminar for a group of astrophysicists. No one quarreled seriously with any of my scientific statements—which again shows the ineffectiveness of peer review—yet more than one told me bluntly that I should keep my doubts to myself. To a man, the rest of those present remained silent on this point. I assumed their silence betokened consent—that to voice misgivings in public was irresponsible. I left the seminar with a bad taste in my mouth. There were now two truths, a truth for scientists and a truth for public consumption. They were both based on the same set of facts but otherwise nonintersecting. Later that day, for the first time in my life, I was called a fascist. I raised scientific questions about a theory that left-wing liberals are morally obligated to support, therefore I was a fascist. Although the remark was made partly in jest, I discovered I did not like it. The same day, a very right-wing physicist congratulated me for doing "God's work." This remark was also made partly in jest, was not put as offensively, but made no more sense.

The thought that a scientist cannot criticize a scientific theory on its scientific merits is more than annoying to a scientist. I have to admit that in my case it is very nearly enough to make me quit science altogether. "There is no such thing as a bad electron," Steven Weinberg is supposed to have said. Apparently he was mistaken. It was also Weinberg who bumped into me in the corridor a few days after the seminar and said that he "enjoyed" my remarks. He thought my comments reasonable but asked if I wouldn't lie to prevent a nuclear war. I replied certainly, if I thought I was not going to be found out. "Of course. That's why there is no point lying in this case, because somebody will sooner or later open his mouth." Over the next few days, more and more of the participants in the seminar approached me and acknowledged that they had been horrified at the suggestion that scientific criticism of nuclear winter be stifled. I had misconstrued their silence. Yet silent they had been. Was this the result of a high moral dilemma on their minds? No, lunch hour was over, the discussion had gone on too long and it was time to get back to classes. This silence betokened life by the clock.

In his book *Weapons and Hope*, Dyson follows George Kennan in the search for a concept of weaponry that is robust: "Above all, a concept should be robust; robust enough to survive mistranslations into various languages, to survive distortion by political pressures and interservice rivalries, to survive drowning in floods of emotion engendered by international crises and catastrophes." I propose that the same criteria should apply to nuclear winter and to any issue of similar importance.

To remain silent is not robust because, as Thomas More learned, silence can betoken many things. It can be turned around and be made to speak loudly. Silence is not robust because "the truth will out." This was Weinberg's point and he had been proven correct before the remark was made. During the last half of 1984, the pages of *Nature* have been filled with an acrimonious debate over nuclear winter. TTAPS have had their say, as have Edward Teller and many others. Now we arrive at the most prosaic silence: the silence of fear. Scientists like my Texas friends are afraid to speak out because they do not know how much has been said already. It is too late to remain silent.

The citation of experts in your favor who in fact disagree with you is not robust because someone will notice before long.

Overstatement is not robust. We have all worried about a nuclear winter backfire. Dyson is correct when he notes that for a backfire to occur "it is not necessary that the theory be proved flatly wrong. It could also happen that the theory is proved to be right but rather simple changes in weapon deployments and targeting rules will be sufficient to make the major effects of nuclear winter disappear." He recalls that Pauling's disarmament campaign collapsed when the Limited Test Ban Treaty was signed in 1963. The campaign was directed against the fallout, not the weapons. A simple technical fix—hiding the weapons underground—removed the threat of fallout but not the threat of nuclear war. "The wave of moral outrage that Carl Sagan has created must be directed at the evil of nuclear war itself and not merely against its consequences."

Although Dyson probably does not know it, his concern is fast becoming reality. We have already seen documents that tell us how to avoid nuclear winter. Theodore A. Postol, summarizing a paper on "Possible Military and Strategic Implications of Nuclear Winter," writes:

> By pointing to these structural problems in our policy, I have tried to explain why a threat of nuclear winter does not necessarily pose planning problems that are significantly more problematic than those we already have. . . .
>
> I believe I have also shown how certain other nuclear effects could be used to construct a nuclear force that can credibly be used to threaten Soviet society with complete destruc-

tion, without causing, in all probability, a nuclear winter effect.

Vice Admiral J. A. Lyons writes in *Science*: "In the long term, the [results] deserve serious study to see what, if any, changes in U.S. targeting policy are required." Let us see if we can make nuclear winter go away without reducing the size of the arsenals.

Nuclear winter has also been considered a weapon of the most perverse sort. Dr. Richard Wagner, Assistant to the Secretary of Defense for Atomic Energy, is quoted in the *New York Times*: "An attacker might be tempted to strike first, right up to the threshold of nuclear winter knowing that 'the prospect of going over the threshold would inhibit the response.' " Such speculations verge on the lunatic. I am gratified to learn that Wagner had at least been "sitting here all morning wondering whether to introduce the thought."

I hope these examples illustrate that not only do planners see nuclear winter as vulnerable to a technical fix, but that it is naive to expect the response will simply be the laying-down of arms by the United States and the Soviet Union.

Carl Sagan has recently published a "one-year anniversary" article in *Parade* entitled "We Can Prevent Nuclear Winter." He has fallen squarely into Dyson's trap: he focuses entirely on the symptoms and not on the disease. Nuclear war is no longer of sufficient consequence to figure in the title. "Quite apart from the radioactivity," Sagan writes, "the toxic smogs, the later enhancement of ultraviolet light from the sun and other effects, it is clear that if the lights go out and the temperature plunges—for months, if not for years—our global civilization and the human species will be profoundly imperiled." The question, whether there will be a global civilization left to imperil, is not asked. Surely this approach, in the profoundest sense, is not robust.

If Dyson will not agree that honesty is better than silence, perhaps he will agree that it is more robust. It has a long history of being the best policy.

In writing this essay I have endeavored to be precise, accurate, and fair. It seems to me, in the long run, to be the simplest and most robust strategy. I embarked on my chronicle largely to clarify the issues for myself. At the end of my chronicle I find that the issues are not so complex after all.

On the scientific level, the First Commandment has clearly been violated. The First Commandment of science states "Thou shalt not believe thine own theories too much" or, more succinctly, "Thou shalt not covet thine own hypotheses." There is no conflict between advocating a hypothesis, and even championing it, and admitting it might be wrong. There is no credit lost in giving someone else his due.

On the medical level, attention has been diverted from the disease of nuclear war to the symptom of nuclear winter. Fixing the one will not cure the other.

On the publicity level, attempts to ensure robustness have undermined it. For a theory to be correct does not require that every scientist in the world sign his name to it. Reality is, I would like to think, not determined by referendum.

On a human level, more than one scientist has convinced himself or herself there is a conflict between saying in public, "I am morally opposed to nuclear war on the grounds that it constitutes mass murder on an unimaginable scale," and "I am not convinced of the nuclear winter calculations but they are worth pursuing." I do not believe the issues are so complex that the public cannot understand the difference between these two statements.

When the air finally clears, we will find the climatologists going about their business: they will argue and they will fight and they will attempt to get closer to the truth. They will do this regardless of what has been said in public, and when the public words have been long forgotten, it will be the results that remain. And if the results should fail to predict with certainty a climatic catastrophe, this is not so important. It is not important because the trust about nuclear war is horrible enough not to require distortion. If the horror is not sufficient to prevent the waging of such a war, then we have already lost the ability to comprehend what will follow. If reason has any chance, there is no need to swerve from the simplest course: say what you know with force and conviction, admit what you don't know, and forget the rest.

POSTSCRIPT

This memoir caused ripples before publication. Several drafts of the manuscript were circulated to the people I cited, and they were given the opportunity to correct any quotations I attributed to them. Apart from the removal of some then-confidential remarks, which mostly took place at a meeting with Soviet scientists, there were no corrections to quotations.

Paul Crutzen saw the piece and wrote to me, "I would be grateful to obtain a copy of your article 'A Memoir of Nuclear Winter,' parts of which I enjoyed reading while I was visiting Rich Turco." A beautifully ambiguous statement if there ever was one. Crutzen said that "I may want to comment on your manuscript at a later date" but never did.

Turco had a number of technical corrections which I have attempted to incorporate into this draft. He may still object to several points and is welcome to respond. Among his major criticisms was that my entire argument falls apart if the number of outright fatalities in a nuclear exchange is much less than the 600 million lower bound I claimed. Indeed, Pentagon estimates are chronically much lower than this—so low that nobody believes them. I do not subscribe with confidence to any such estimates, which I basically regard as slightly refined guesswork. My numbers represent an average compiled from various sources. Even Sagan in his *Parade* article says that "more than 2 billion—almost half of all the humans on earth—would be destroyed in the immediate aftermath of a thermonuclear war." Two billion is enough for me.

In regard to the 100-megaton "threshold" scenario, this is a somewhat tricky matter and deserves more detailed comment. The strictly correct statement is that the TTAPS code does not target cities at all; since it is a one-dimensional code, it cannot do so. It merely computes a total amount of smoke from the formula $S = A \times f \times e$, where A is the area burned, f is the fuel burned per unit area, and e is the number of grams of smoke emitted per gram of fuel burned. One then has to go to obscure demographic sources to decide how many cities you need to torch to produce this amount of smoke. Furthermore, you see clearly that if you lower A, the area burned, you can keep S the same by raising f or e. This is essentially what TTAPS have done in the 100-megaton scenario and their result gives about 60% of the baseline smoke. You also find that 1,000 city centers are needed to produce the reported results.

That's really all there is to it, but for the technically oriented reader I can be more specific. The smoke (in grams) produced by cities for the baseline case was computed as

$$S_b = 250,000\,(.95 \times 3 \times .027 + .05 \times 10 \times .011) \times 10^{10}.$$

Here, the 250,000 is the total city area targeted by the baseline scenario in square kilometers, and the 10^{10} factor at the far right is to convert this to square centimeters. Of this 250,000 square kilometers, TTAPS assumes that 95% (.95) is composed of "suburban" areas with a fuel loading of $f = 3$ grams/cm^2 and an emission factor of $e = .027$. Five percent of the city area (.05) is assumed to be "urban" with a fuel loading of $f = 10$ grams/cm^2 and an emission factor of $e = .011$. One sees that the suburban areas generate about 119×10^{12} grams of smoke while the urban areas generate only about 14×10^{12}, roughly ten times less.* Adding in forest smoke brings the total for the baseline case up to about 225×10^{12} grams.

Now, the 100-megaton city scenario ignores forests and suburban areas and targets only the heavily fuel-laden urban centers. The parameters used are

$$S_c = 25,000\,(1 \times 20 \times .026) \times 10^{10}.$$

While the area burned has been decreased by a factor of ten, we see that the fuel loading f has been doubled from the 10 grams/cm^2 used in the baseline case, and the emission factor raised from $e = .011$ to $e = .026$. The total smoke, S_c, is now about 130×10^{12} grams, more than that produced by cities in the baseline exchange. It is also about 60% of the baseline smoke when forests are included in the baseline calculation.

One must now ask how many urban centers give you 25,000

---

* The two or three readers who may have followed the nuclear winter calculations closely know that the numbers actually used in the TTAPS calculations vary from the published values. The numbers here are the values published in *Science*.

square kilometers at a fuel loading of 20 grams/cm^2. This sort of data you dig up in atlases and the like and, as mentioned above, the answer is roughly 1,000, not 100 as claimed by Sagan. You might respond that 100 cites with a fuel loading of 200 grams/cm^2 will give you the same answer. This is true, but according to the limited amount of existing data (the same data TTAPS used), city centers with a fuel burden of 200 grams/cm^2 do not exist. This is evidently why TTAPS assumed that the 100 megatons were distributed in 1,000 explosions and not 100.

On other matters, since 1984 much work has been done on nuclear winter, and the studies I focused on in this memoir are essentially obsolete. Three-dimensional, interactive calculations have been pursued by MacCracken and Covey, Schneider and Thompson. The effects of oceans, infrared absorption and particle coagulation, among other things, have now been modeled. Not surprisingly, all these details tend to mitigate the original TTAPS results.

In the summer 1986 issue of *Foreign Affairs*, Thompson and Schneider write, ". . . on scientific grounds the global apocalyptic conclusions of the initial nuclear winter hypothesis can now be relegated to a vanishingly low level of probability." Currently popular to describe the results is "nuclear fall," a term Barbara Levi and I used at our initial meeting with Soviet scientists during December 1983 and which angered both Aleksandrov and Evgeny Velikov, vice-president of the Soviet Academy of Sciences.

Nevertheless, the statement of Thompson and Schneider strikes me as a miniature example of the backfire I discussed in the memoir. The results they present still show temperature drops over land of about 10°C for a period of days to weeks, perhaps enough to cause another "year without a summer." Had the initial predictions not been so dramatic, this might still be considered a major disaster.

BIBLIOGRAPHY

Because this was a somewhat anecdotal memoir, I did not want to clutter it up with footnotes. I here list my sources.

1

Carl Sagan, "The Nuclear Winter," *Parade* (October 30, 1983), p. 4.

2

Paul J. Crutzen and John W. Birks, "The Atmosphere After a Nuclear War: Twilight at Noon," in *The Aftermath* (New York: Pantheon Books, 1983); Richard Turco et al., "Nuclear Winter: Global Consequences of Multiple Nuclear Weapons Explosions," *Science* 222 (1983): 1283; Paul R. Ehrlich et al., "Long-Term Biological Consequences of Nuclear War," *Science* 222 (1983): 1293.

The last two papers are reprinted as appendixes in Paul R. Ehrlich and Carl Sagan, eds., *The Cold and the Dark* (New York: W. W. Norton, 1984).

Other papers along similar lines are: Paul R. Ehrlich, "The Nuclear Winter," *The Amicus Journal* (Winter 1984), p. 19; Richard Turco et al., "The Climatic Effects of Nuclear War," *Scientific American* 251 (1984): 33.

The initial confirmatory studies were: Michael C. MacCracken, "Nuclear War: Preliminary Estimates of the Climatic Effects of a Nuclear Exchange," Lawrence Livermore Laboratory preprint, UCRL-89770 (October 1983); V. V. Aleksandrov and G. L. Stenchikov, "On the Modelling of the Climatic Consequences of Nuclear War," USSR Academy of Sciences Computing Centre, Moscow (1983); Curt Covey et al., "Global Atmospheric of Nuclear War, etc.," *Nature* 308, (1984): 21.

The prehistoric nuclear winter studies were: Tom Stonier, *Nuclear Disaster* (Cleveland: World Publishing Co., 1963); E. S. Batten, "The Effects of Nuclear War on the Weather and Climate," Rand Corporation, Memorandum RM-4989-Tab (1966); R. U. Ayres, "Environmental Effects of Nuclear Weapons," Hudson Institute, Inc., HI 518 (1965); W. J. Humphreys, *Physics of the Air*, 3d ed. (New York: McGraw-Hill, 1940).

The quotations from Kissinger, York, and Wiesner and the Soviet government are as found in: Alva Myrdal, *The Game of Dis-*

armament (New York: Pantheon Books, 1982), pp. 29 and 38; David Holloway, *The Soviet Union and the Arms Race* (New Haven: Yale University Press, 1984), p. 42.

3

Aleksandrov's discussion of his results can be found on pp. 95–100 of *The Cold and the Dark* (see Section 2, above). Kapitza's statement is simply entitled "Prepared Statement" and is in my possession.

4

Sagan's presentation of the "threshold" scenario can be found in *The Cold and the Dark* (see Section 2, above), pp. 19–20. The diagrams I refer to are on p. 17.

Sagan's citation of the confirmatory studies is from *Foreign Affairs* (Winter 1984), p. 264.

The "harsh, if accurate," criticism of the Soviet study from Thompson and Turco can be found in *Science* 225 (1984): 31. See also pp. 978–979 of the same volume.

The full title of the NAS report is, "The Effects on the Atmosphere of a Major Nuclear Exchange" (Washington, D.C.: National Academy Press, 1985), pp. 1–7.

5

The TTAPS rebuttal is found in Richard Turco et al., *Nature* 311 (1984): 307; the lengthy Ehrlich passage is from his *Amicus* article (see Section 2, above), pp. 23–24; Carrier's remark is quoted by *Nature* 312 (1984): 683; and for Turco's comment, see the *New York Times*, August 5, 1984, p. 40.

6

Robert Bolt, *A Man for All Seasons* (New York: Vintage Books, 1962), p. 88; Freeman Dyson, "Star Wars, Austrianization and

Nuclear Winter,'' originally given as the Albert Pick Lecture at the University of Chicago (April 4, 1985); Freeman Dyson, *Weapons and Hope* (New York: Harper and Row, 1984), p. 226; Theodore A. Postol, ''Possible Military and Strategic Implications of Nuclear Winter,'' Center for International Security and Arms Control, Stanford University (September 12, 1984); Carl Sagan, ''We Can Prevent Nuclear Winter,'' *Parade* (September 30, 1984).

Admiral Lyon's remarks appeared in *Science* 225 (1984): 31; and Wagner's remarks are as reported in the *New York Times*, August 5, 1984, p. 40.

POSTSCRIPT

Starley L. Thompson and Stephen H. Schneider, ''Nuclear Winter Reappraised,'' *Foreign Affairs* (Summer 1986), p. 981; Starley L. Thompson, ''Global Interactive Transport Simulations of Nuclear War Smoke,'' *Nature* 317 (1985): 35; Michael C. MacCracken and John J. Walton, ''The Effects of Interactive Transport and Scavenging of Smoke on the Calculated Temperature Change etc.,'' Lawrence Livermore Laboratory preprint, UCRL-91446 (December 1984).

6·GENIUS AND BIOGRAPHERS: THE FICTIONALIZATION OF EVARISTE GALOIS

1. INTRODUCTION

In Paris, on the obscure morning of May 30, 1832, near a pond not far from the pension Sieur Faultrier, Evariste Galois confronted an adversary in a duel to be fought with pistols, and was shot through the stomach. Hours later, lying wounded and alone, Galois was found by a passing peasant. He was taken to the Hôpital Cochin, where he died the following day in the arms of his brother Alfred, after having refused the services of a priest. Had Galois lived another five months, until October 25, he would have attained the age of twenty-one.

The legend of Evariste Galois, one of the creators of group theory, has fired the imagination of generations of mathematics students. Many of us have experienced the excitement of Freeman Dyson, who writes:

> In those days, my head was full of the romantic prose of E. T. Bell's *Men of Mathematics*, a collection of biographies of the great mathematicians. This is a splendid book for a young boy to read (unfortunately, there is not much in it to inspire a girl, with Sonya Kovalevsky allotted only half a chapter), and it has awakened many people of my generation to the beauties of mathematics. The most memorable chapter is called "Genius and Stupidity" and describes the life and

death of the French mathematician Galois, who was killed in a duel at the age of twenty.[1]

Dyson goes on to quote Bell's famous description of Galois' last night before the duel:

All night long he had spent the fleeting hours feverishly dashing off his scientific last will and testament, writing against time to glean a few of the great things in his teeming mind before the death he saw could overtake him. Time after time he broke off to scribble in the margin "I have not time; I have not time," and passed on to the next frantically scrawled outline. What he wrote in those last desperate hours before the dawn will keep generations of mathematicians busy for hundreds of years. He had found, once and for all, the true solution of a riddle which had tormented mathematicians for centuries: under what circumstances can an equation be solved? [2]

This extract is likely the very paragraph which has given the greatest impetus to the Galois legend. As with all legends the truth has become one of many threads in the embroidery. E. T. Bell has embroidered more than most but he is not alone. James R. Newman, writing in *The World of Mathematics*, notes: "The term *group* was first used in a technical sense by the French mathematician Evariste Galois in 1830. He wrote his brilliant paper on the subject at the age of twenty, the night before he was killed in a stupid duel."[3] From the prospectus of the famed Bullitt archives of mathematics issued by the University of Louisville library, we learn: "Goaded by a 'mignonne' and two of her slattern confederates into a 'duel of honor,' Galois was shot and killed at the age of twenty."[4] In John Sommerfield's novel *The Adversaries*, based on Bell's account, Galois is given the name Roger Constant and "much fuss is made about the woman."[5] Leopold Infeld, in his biography of Galois,[6] invokes a conspiracy theory to explain Galois' death: Galois was considered one of the most dangerous republicans in Paris; the government wanted to get rid of him; a female agent provocateur set him up for the duel; et cetera. Fred

Hoyle, in his *Ten Faces of the Universe*,[7] attempts a partial inversion of the argument: Galois' ability to carry on complex calculations entirely in his head made him appear distant to others; personal animosities arose with republican friends; they began to think he was not entirely for the cause; Galois in their eyes was the agent provocateur; et cetera. All three authors, Bell, Hoyle, and Infeld, invoke a political cause for the duel, with a mysterious coquette just off center.

This essay is an attempt to sift some of the facts of Galois' life from the embroidery. It will not be an entirely complete account and will assume the reader is familiar with the story, presumably through Bell's version. Because these authors have emphasized the end of Galois' life, I will do so here. As will become apparent, many of the statements just cited are at worst nonsensical, and at best have no basis in the known facts.

Although a number of the documents presented here are, I believe, translated into English for the first time, it should be emphasized that they are not new, just ignored. There is more known about Galois than recent authors admit. In the first version of the essay, which appeared in the *American Mathematical Monthly*, I expressed the hope that some ambitious historian would find the requisite letter in an attic trunk or a newspaper clipping in the Paris archives to unravel the remaining mysteries. As it turned out, unbeknownst to me, the clipping had already been found a number of years ago. Remaining mysteries are now very few. Those of you who prefer to bask in the warm orange glow of the mysterious should stop now; the rest of you can carry on.

2. SOURCES

It is not difficult to trace the story of Galois' brief life through its increasingly embellished incarnations. The primary source of information, containing eyewitness accounts and many relevant documents, is the original study of Paul Dupuy, which appeared in 1896.[8] Dupuy was a historian and the Surveillant Général of the Ecole Normale. Bell, Hoyle, and Infeld all cite it as an important reference but never once explicitly quote it. Indeed, Bell acknowledges that his account is based on Dupuy[9] and the documents in Tannery (see below); but it remains unclear how much of his forebearer Bell has read, for while numerous passages are lifted bodily from Dupuy, other important information contained in the Surveillant Général's biography is strangely absent. Dupuy's study itself is lacking a number of important letters and documents. Whether Dupuy was unaware of their existence or chose not to publish them I do not know. He also makes a number of minor errors in chronology. In any case, the first lesson is already learned: those who use Dupuy as their sole source of information must make mistakes. Nevertheless, this original biography is much more complete and accurate than subsequent dilutions and contains more information than a reading of Bell, Hoyle, or Infeld would even suggest. A translation of Dupuy into English should be undertaken.

Some of the documents not found in Dupuy are contained in Tannery's 1908 edition of Galois' papers.[10] All can be found in the definitive 1962 edition of Bourgne and Azra.[11] This volume contains every scrap of paper known to have been written by Galois, an accurate chronology, facsimiles of some of his original manuscripts, and a number of relevant letters by others. When quoting Galois, I have worked exclusively from this edition.

The memoirs of Alexandre Dumas[12] contain a pertinent chapter, and the *Lettres sur les prisons du Paris* by François Vincent Raspail[13] are the primary source on Galois' months in prison. Some of these letters are quoted by Dupuy and Infeld. Other references will be cited as they appear.

3. EARLY LIFE AND LOUIS-LE-GRAND

I will not dwell at length on the first sixteen years of Galois' life, for they are reported with fair accuracy by Bell. This is not surprising; his account approaches that of a somewhat abridged translation of Dupuy. The divergences will set in later. Hence this section and the next may be taken as a rather condensed review and criticism of Bell. Infeld and Hoyle, who concentrate most of their energies on the duel, will be dealt with at the appropriate time.

Evariste Galois was born on October 25, 1811, not far from Paris in the town of Bourg-la-Reine, France. His father was Nicholas-Gabriel Galois, who was then thirty-six years of age, and his mother was Adelaide-Marie Demante. Both parents were highly intelligent and well educated in the subjects considered important at the time: philosophy, classical literature, and religion. Bell points out that there is no record of mathematical talent on either side of the family. A more neutral statement should perhaps be made: no record exists in favor of or against any such talent. M. Galois did possess the gift for composing rhymed couplets with which he would amuse neighbors. This harmless activity, as Bell notes, would later cause his undoing. Evariste seems to have inherited some of his ability, participating in the fun at house parties. For the first twelve years of his life, Evariste's mother served as his sole teacher, giving him a solid background in Greek and Latin, as well as passing on her own skepticism toward religion.

In 1815, during the One Hundred Days, M. Galois was elected mayor of Bourg-la-Reine. He had been a supporter of Napoleon and, in fact, had been elected chief of the town's liberal party during Napoleon's first exile. After Waterloo, he had planned to relinquish his post to his predecessor, but the latter had left the country. M. Galois demanded to be either confirmed or replaced, and in the confusion managed to keep his office. He served the new king faithfully, but from this point on he met increasing resistance from the conservative elements of his town. It is probably safe to say that the younger Galois inherited his liberal ideas from his parents.

On October 6, 1823, Evariste Galois was enrolled in the Lycée de Louis-le-Grand, a famous preparatory school (which still exists) in Paris.* Both Robespierre and Victor Hugo had studied there. Louis-le-Grand is where his troubles began, where Infeld's account of Galois' life essentially opens, and where Bell introduces his theme of "Genius and Stupidity," taking on the tone of a blanket condemnation of almost everyone and everything that surrounded Galois. "Galois was no 'ineffectual angel,' " Bell

---

* At the time the school was called the Collège de Louis-le-Grand.

writes in his introduction, "but even his magnificent powers were shattered before the massed stupidity aligned against him, and he beat out his life fighting one unconquerable fool after another."[14] I believe we will see that the problems ran much deeper than that.

Bell's first liberties with Dupuy are minor. Bell describes Louis-le-Grand as a "dismal horror" and goes on to say "the place looked like a prison and was."[15] Admittedly, Dupuy writes that Louis-le-Grand happened to look like a jail because of its grills, but he then goes on to describe the underlying "passions of work, academic triumph, passions of liberal ideals, passions of memories of the Revolution and Empire, contempt and hate for the legitimist reaction."[16] Bell, by cutting Dupuy's sentence in half, has begun the slant to the negative.

At this particular time, there were undeniable problems. During Galois' first term, the students, who suspected the new provisor of planning to return the conservative Jesuits to the school, protested by staging a minor rebellion. When required to sing at a chapel service, they refused. When required to recite in class, they refused. When required to toast Louis XVIII at an official school banquet, they refused. The provisor summarily expelled the forty students whom he suspected of leading the insurrection. Galois was not among those expelled, nor is it known if he was even among the rebels, but we may guess that the arbitrariness of the provisor and the general severity of the school's regime made a deep impression on him.

Nevertheless, Galois' first two years at Louis-le-Grand were marked by a number of successes. He received the first prize for Latin verse and three honorable mentions, as well as a mention in Greek for the General Concourse. At this point we witness the first of Bell's distortions of chronology to give the impression that Galois was misunderstood and persecuted. Galois was asked to repeat his third year because of his poor work in rhetoric. Bell writes, "His mathematical genius was already stirring," and "He was forced to lick up the stale leavings which his genius had rejected."[17] I cannot say with any certainty whether Galois' mathematical genius was already stirring, but it is known that Evariste did not enroll in his first mathematics course until *after* he had been demoted.

During his mathematics course, which he began in February 1827, Galois discovered Legendre's text on geometry, soon followed by Lagrange's original memoirs: *Resolution of Numerical Equations,* Theory of Analytic Functions* and *Lessons on the Calculus of Functions.* Doubtlessly, Galois received his initial ideas on the theory of equations from Lagrange. I do not understand why Bell claims Galois' classwork was mediocre; his instructor M. Vernier constantly writes such accolades as "zeal and success," "zeal and progress very marked."[18]

With his discovery of mathematics, Galois became absorbed and neglected his other courses. Before enrolling in M. Vernier's class, typical comments about him had been:[19]

Religious Duties—Good	Work—Sustained
Conduct—Good	Progress—Marked
Disposition—Happy	Character—Good, but singular

After a trimester in M. Vernier's class, the comments were:

Religious Duties—Good	Work—Inconstant
Conduct—Passable	Progress—Not very satisfactory
Disposition—Happy	Character—Withdrawn and original

The words "singular," "bizarre," "original," and "withdrawn" would appear more and more frequently during the course of Galois' career at Louis-le-Grand. His own family began to think him strange. His rhetoric teachers would term him "dissipated." Bell discusses these remarks at some length and his use of the indefinite pronoun "they" gives the impression that the entire faculty was aligned against Galois. A perusal of Dupuy's appendix, however, shows the negative remarks were penned, by and large, by Galois' two rhetoric teachers. Until this point in his life,

---

* Here and elsewhere, read "algebraic equations" for "numerical equations."

I believe it is fair to say that Galois was somewhat misunderstood by his teachers in the humanities, but not that he was persecuted.

Slightly more serious problems were soon to arise. His mathematics teacher, M. Vernier, constantly implored Galois to work more systematically. His remark on one of Galois' trimester reports makes this clear: "Intelligence, marked progress, but not enough method."[20] Galois did not take the advice; he took the entrance examination to l'Ecole Polytechnique a year early, without the usual special course in mathematics, and failed. To Galois, his failure was a complete denial of justice. This and subsequent rejections embittered him for life. When we examine some of his later writings, I think it will be evident that he developed not a little paranoia.

Galois did not yet give up. The same year, 1828, saw him enroll in the course of Louis-Paul-Emile Richard, a distinguished instructor of mathematics. Richard encouraged Galois immensely, even proclaimed that he should be admitted to the Polytechnique without examination. The results of such encouragement were spectacular. In April 1929, Galois published his first short paper, "Proof of a Theorem on Periodic Continued Fractions." It appeared in the *Annales de Gergonne*.

This paper was a minor aside. Galois had also been working on the theory of equations ("Galois theory"). On May 25 and June 1, 1829, while still only seventeen, he submitted to the Academy his first researches on the solubility of equations of prime degree. Cauchy was appointed referee.

We now encounter a major myth which evidently had its origins in the very first writings on Galois and which had been perpetuated by virtually all writers since. This myth is the assertion that Cauchy either forgot or lost the papers (Dupuy, Bell[21]) or intentionally threw them out (Infeld[22]). Recently, however, René Taton has discovered a letter of Cauchy's in the Academy archives which conclusively proves that he did not lose Galois' memoirs but had planned to present them to the Academy in January 1830.[23] There is even some evidence that Cauchy encouraged Galois. I will discuss this letter and related events in more detail below; for now I note only that to hold Cauchy responsible for "one of the major disasters in the history of mathematics," to paraphrase Bell,[24] is simply incorrect, and to add neglect by the Academy to the list of Galois' difficulties during this period appears entirely unwarranted.

A truly tragic blow came within a month of the submission: on July 2, 1829, Galois' father committed suicide. The reactionary priest of Bourg-la-Reine had signed Mayor Galois' name to a number of maliciously forged epigrams directed at Galois' own relatives. A scandal erupted. M. Galois' good nature could not stand such an attack and he asphyxiated himself in his Paris apartment "not two steps from Louis-le-Grand." During the funeral, when the same clergyman attempted to participate, a small riot erupted. The loss of his father may explain much of Galois' future behavior. We must wait a few years, until his second prison term, to see him. In any case, he loved his father dearly, and if an iron link had not already been forged between the Bourbon government and the Jesuits, it had now.*

But Galois' troubles were not yet over. A few days later, he failed his examination to the Ecole Polytechnique for the second and final time. Legend has it that Galois, who worked almost entirely in his head and who was poor at presenting his ideas verbally, became so enraged at the stupidity of his examiner that he hurled an eraser at him. Bell records this as a fact,[25] but according to the little-known study of Joseph Bertrand[26] the tradition is false. Bertrand, who appears to have detailed information about the event, records that Galois, while expounding on the properties of logarithmic series, refused to prove his statements to the examiner M. Dinet and, in response to Dinet's questions, replied merely that the answer was completely obvious. So was the result.

The examination failures, as well as the misunderstanding of his humanities teachers, left Galois irrevocably embittered. Bell quotes him as writing, "Genius is condemned by a malicious social organization to an eternal denial of justice in favor of fawning

---

* Probably the clearest picture of the relationship between the Jesuits and the Bourbons, one which contains episodes paralleling that of M. Galois' misfortunes, is Stendhal's famous novel *The Red and the Black*.

mediocrity.''[27] I believe Bell constructed this quotation from a passage of Dupuy's[28] but Galois did express similar sentiments in his fragmentary essay, ''Sciences, Hiérarchie: Ecoles'' and in ''Sur l'Enseignement des sciences'' (''Hierarchy is a means for the inferior.''[29]). In Bell's diatribe against this famous examination, as well as in other accounts of it, the suggestion that the death of Galois' father several days before may have had something to do with the outcome never arises. It is a simple matter for Bell to lay the fault squarely with the examiner's stupidity because Bell has placed the examination before M. Galois' unfortunate suicide. In this instance, Bell is not fully to blame; Dupuy does not date the examination.[30] I do not wish to suggest Galois should have been failed. I only wish to point out that the examination must have been held under the worst possible conditions.

Thus, Galois' secondary-school career ended in a string of minor setbacks and two major disasters. Galois had not planned to take the Baccalaureate examinations, because the Ecole Polytechnique did not require them. Now, having failed the Polytechnique's entrance examination and having decided to enter the less prestigious Ecole Normale,* he was forced to reconsider. ''Still persecuted and maliciously misunderstood by his preceptors,'' in Bell's words, ''Galois prepared himself for the final examinations.''[31] Despite such malice, Galois did very well in mathematics and physics, though less well in literature. He received both a Bachelor of Letters and a Bachelor of Science on December 29, 1829.

It is interesting to note that, although he has continued to play the role of muckraker of malice, Bell fails to mention M. Richard's distinct cooling toward Galois, on whom he had previously bestowed high economia. After the first trimester of the 1828–1829 academic year, Richard wrote: ''The student is markedly superior to all his classmates.'' After the second: ''This student

---

* Then called the Ecole Préparatoire. The faculty of the Polytechnique before 1830 included, among others, Lagrange, Laplace, Fourier, Ampère, Cauchy, Coriolis, Poisson, and Gay-Lussac. It is not surprising that young scientists wished to enroll there.

works only in the highest realms of mathematics." After the third: "Conduct good, work satisfactory."

Because I do not have an accurate date for this report, I cannot propose a specific event as the cause of this obvious change in attitude. Presumably it occurred in the spring of 1829, shortly before or after Galois' time of troubles began. One could of course argue that M. Richard has simply become bored with Galois. Otherwise it does serve to show that Bell's black-and-white presentation of Galois' preceptors is an oversimplification.

4. THE ECOLE NORMALE

The early months of 1830, which saw Galois officially enrolled as a student at the Ecole Normale, also witnessed an interesting series of transactions with the Academy. Recall that Galois had submitted his first researches to the Academy on May 25 and June 1, 1929. On January 18, 1830, Cauchy wrote the previously mentioned letter discovered by Taton:[32]

> I was supposed to present today to the Academy first a report on the work of the young Galois, and second a memoir on the analytic determination of primitive roots in which I show how one can reduce this determination to the solution of numerical equations of which all roots are positive integers. Am indisposed at home. I regret not being able to attend today's session, and I would like you to schedule me for the following session for the two indicated subjects.
>
> Please accept my homage. . . .
>
> A.-L. Cauchy

This letter makes it clear that, six months after their receipt, Cauchy was still in possession of Galois' manuscripts, had read them, and was very likely aware of their importance. At the following session on January 25, however, Cauchy, while presenting his own memoir, did not present Galois' work. Taton hypothesizes that between January 18 and January 25, Cauchy persuaded Galois to combine his researches into a single memoir to be submitted for the Grand Prize in Mathematics, for which the deadline was

March 1. Whether or not Cauchy actually made the suggestion cannot yet be proved, but in February Galois did submit such an entry to Fourier in his capacity of perpetual secretary of mathematics and physics for the Academy. In any event, there is an additional piece of evidence that attests to Cauchy's appreciation of Galois' work. This is an article which appeared the following year on June 15, 1831, in the Saint-Simonian journal *Le Globe*. The occasion was an appeal for Galois' acquittal after his arrest following the celebrated banquet at the restaurant Vendanges des Bourgogne:

> Last year before March 1, M. Galois gave to the secretary of the Institute a memoir on the solution of numerical equations. This memoir should have been entered in the competition for the Grand Prize in Mathematics. It deserved the prize, for it could resolve some difficulties that Lagrange had failed to do. *M. Cauchy had conferred the highest praise on the author about this subject.* And what happened? The memoir is lost and the prize is given without the participation of the young *savant*. [Taton's italics]*[33]

The misfortune referred to above was the death of Fourier on May 16, 1830. Galois' entry could not be found among Fourier's papers and in Galois' eyes this could not be an accident. "The loss of my memoir is a very simple matter," he wrote. "It was with M. Fourier, who was supposed to have read it and, at the death of this *savant*, the memoir was lost."[34] It was an unfortunate coincidence; however, it was not Fourier's sole responsibility to read the manuscript, for the committee appointed to judge the

* My own interpretation of this article is slightly different than that of Taton. Taton writes that the journalist evidently had firsthand information. But note the date: June 15, 1831. In the aftermath of the July revolution, Cauchy fled France in September, almost nine months prior to the article's publication. It is difficult to see when the journalist would have spoken to Cauchy. However, the article appeared in a Saint-Simonian journal. Galois' best friend, Auguste Chevalier, was one of the most active Saint-Simonians. My own suspicion, which I cannot prove, is that the journalist was Chevalier and the information was coming directly from Galois. If this hypothesis is correct, Galois himself is admitting Cauchy's encouragement.

Grand Prize consisted also of Lacroix, Poisson, Legendre, and Poinsot.[35] I mention this because a number of sources give the impression that somehow Fourier either intentionally "lost" the paper or could not understand it.[36]

In spite of the setback caused by this loss of his manuscript, Galois' published his paper "An Analysis of a Memoir on the Algebraic Resolution of Equations" in the April issue of the *Bulletin de Ferussac*. In June he published "Notes on the Resolution of Numerical Equations" and the important article "On the Theory of Numbers."[37]

In addition to propagating the legend that Cauchy lost the manuscripts, Bell, curiously, does not mention Fourier by name in the preceding misadventure, although Dupuy is explicit on the identity of the Academy's Perpetual Secretary. Perhaps Bell felt it a little too much to "expose" Cauchy, Fourier, and later Poisson as incompetents. Bell also does not make it clear that the papers listed above (plus a later memoir) constitute what is now called Galois theory. If this point had been clarified, the claim that Galois had written down the theory on the eve of the duel would be difficult to substantiate or even to suggest.

From this point onward, the scenario of Galois as a passive victim to negligence, misunderstanding, and bad luck begins to break down—if it has not already. More and more he participated in the creation of his own disasters. But this picture does not fit Bell's plan. Therefore chronology is rearranged, events are omitted, and others are invented in increasing quantity, until the end of his account is largely fantasy. The wholesale reordering of events will be especially evident in what follows.

Most important, Bell gives an extremely late start to Galois' political activities. He remarks that had Evariste's teachers at Louis-le-Grand allowed him to study only mathematics, he might have lived to be eighty.[38] Unlikely. According to Dupuy, one of the reasons Galois had hoped to attend the Polytechnique was to participate in political activities. At the Ecole Normale he became a "polytechnician in exile." The July revolution of 1830 reared its head. The director of the school, M. Guigniault, locked the students in so that they would not be able to fight on the streets.

Galois was so incensed at the decision that he tried to escape by scaling the walls. He failed, and in doing so missed the revolution. Afterwards, the director put the students in the service of the provisional government. Charles X had fled France. He would be followed in September by Cauchy. Louis-Philippe was the new king.

The events of July, severely abridged here, Bell chronicles accurately. He does fail to point out that Galois probably joined the Society of Friends of the People, one of the most extreme republican secret societies, within the next month, certainly before December.[39] I will explain the importance of this omission after I fill in the remaining gaps of the narrative as they took place historically.

In December of that year, M. Guigniault was engaging in polemics against students in the pages of several newspapers. Galois saw his chance for attack and jumped into the squabble with a blistering letter to the *Gazette des Ecoles*. It read in part:

Gentlemen:

The letter which M. Guigniault inserted yesterday in the *Lycée* on the occasion of one of the articles in your journal has seemed to me very inappropriate. I had thought that you would welcome with eagerness every means to expose this man.

Here are the facts which can be verified by forty-six students.

On the morning of July 28, when many of the students wished to leave the school and fight, M. Guigniault told them on two occasions that he would call the police to reestablish order within the school. The police on the 28th of July!

On the same day, M. Guigniault told us with his usual pedantry: "There are many brave men fighting on both sides. If I were a soldier I would not know what to decide—to sacrifice liberty or LEGITIMACY?"

Here is the man who the next day covered his hat with an immense tricolor cockade. Here are our liberal doctrines![40]

Galois continues. According to Dupuy, every statement in the letter is accurate. Nevertheless, the result is what might have been anticipated: Galois was expelled. The action was to become official on January 4, but Galois quit school immediately and joined the Artillery of the National Guard, a branch of the militia which was almost entirely composed of republicans. It is interesting to note that the forty-six students referred to in the letter actually published a reply *against* Galois, but this seems to have been at the "prompting" of M. Guigniault.[41]

December was a turbulent month for other reasons. After the Bourbons had fled France, four of their ex-ministers were to be tried for treason. Popular sentiment called for their execution. The decision to execute or imprison for life was to be announced on December 21. That day, the Artillery of the National Guard was stationed in the quadrangle at the Louvre. Galois was certainly there. The atmosphere was very tense. If the ministers were given a life sentence, the artillerymen had planned to revolt. But the Louvre was soon surrounded by the full National Guard and troops of the line, the more trustworthy arms of the military. A distant cannon shot was heard. It signaled the end of the trial and that the ministers had indeed been given imprisonment over execution. The artillerymen and the National Guard readied themselves for bloodshed, but with the arrival at the Louvre of thousands of Parisians, the fighting did not erupt. Over the next few days, the situation in Paris grew calmer with the appearance of Lafayette, who called for peace, and daily proclamations calling for order. On December 31, 1830 the Artillery of the National Guard was abolished by a royal decree in fear of its threat to the throne.[42]

In January 1831 Galois, no longer a student, attempted to organize a private class in algebra. At the first meeting about forty students appeared,[43] but the endeavor did not last long, evidently because of Galois' political activities. On the seventeenth of that month, upon the invitation of Poisson, Galois submitted a third version of his memoir to the Academy. Later, in July, Poisson would reject the manuscript.* This rejection will be discussed at

---

* Dupuy does not date this event and its placement in the narrative may have misled Bell.

the proper time, but we should note that by that time Galois would have already been arrested.

If we now leave history to return to Bell's account, we find a totally distorted chain of events: The months after July are missing; Galois still has not joined the Society of Friends of the People. He leaves school in December but has not joined the artillery. The events at the Louvre, which will turn out to have critical importance for the remainder of the story, never take place. Galois attempts to organize his private course in mathematics and, in Bell's words, "Here he was at nineteen, a creative mathematician of the first rank, peddling to no takers. . . . Finding no students, Galois temporarily abandoned mathematics and joined the artillery of the National Guard."[44] According to Bell, Galois submits his paper to Poisson, it is rejected; and this being the "last straw," Galois decides to devote all his energy to revolutionary politics."

The chronology presented by Bell is thus completely backwards. The impression given by this rearrangement of events is once again that of a misunderstood and persecuted Galois who, surrounded on all sides by idiots, finally gives up and goes into radical politics. By writing that Galois found no students, Bell of course strengthens this impression. A more balanced account clearly requires what is lacking in Bell: a Galois of volition. We may get a better indication of his character and behavior during the spring of 1831 from a letter written on April 18 by the mathematician Sophie Germain to her colleague Libri:

Decidely there is a misfortune concerning all that touches upon mathematics. Your preoccupation, that of Cauchy, the death of M. Fourier, have been the final blow for this student Galois who, in spite of his impertinence, showed signs of a clever disposition. All this has done so much that he has been expelled from l'Ecole Normale. He is without money and his mother has very little also. Having returned home, he continued his habit of insult, a sample of which he gave you after your best lecture at the Academy. The poor woman fled her house, leaving just enough for her son to live on, and has been forced to place herself as a companion in order to make ends meet. They say he will go completely mad and I fear this is true.[45]

Unfortunately, as Bell observes, Galois was no ineffectual angel.

Before continuing, another historical detail should be mentioned. As an aftermath of the December events at the Louvre and the dissolution of the Artillery of the National Guard, nineteen officers were arrested, having been suspected of planning to deliver their cannons to the people. The charge was conspiracy to overthrow the government. In April, all nineteen were acquitted.

Until now my criticism has been devoted almost entirely to Bell. Partly this has been because his account is by far the most famous but there are other reasons as well. Hoyle's short essay, as already mentioned, is purely concerned with Galois' death and thus has little to say concerning the foregoing events. Infeld's account, on the other hand, is of book length. In a single article it would be difficult to debate all salient points. Nonetheless, Infeld has also stated that he is primarily concerned with the events surrounding the duel.[46] It is then reasonable to devote attention here to that aspect of the book. Infeld's work is actually something of a curiosity. The bulk of it is a fictionalized biography, interspersed with real documents and eyewitness accounts. All dates, names, and places are respected. The second part of the biography consists of a lengthy Afterword, in which Infeld details exactly what he has invented, what he has not, and what he believes to be true. He also includes a fairly comprehensive bibliography. In my criticisms of Infeld that follow, I take issue only with what he claims not to have invented. The reader may get the flavor of the author's intent by noting that at Galois' private algebra class, spoken of earlier, Infeld has stationed two police spies.[47]

It might also be noted that, according to James R. Newman's brief remark quoted in the Introduction, Galois at this point in the narrative would be dead.

5. ARREST AND PRISON

And thus we arrive on May 9, 1831. The occasion was the republican banquet held at the restaurant Vendanges des Bourgogne, where approximately two hundred republicans were gathered to celebrate the acquittal of the nineteen republicans on conspiracy charges. As Dumas says in his memoirs, "It would be difficult to find in all Paris, two hundred persons more hostile to the government than those to be found reunited at five o'clock in the afternoon in the long hall on the ground floor above the garden."[48] It is worth quoting Bell's description of this event:

> The ninth of May, 1831, marked the beginning of the end. About two hundred young republicans held a banquet to protest against the royal order disbanding the artillery which Galois had joined. Toasts were drunk to the Revolutions of 1789 and 1793, to Robespierre, and to the Revolution of 1830. The whole atmosphere of the gathering was revolutionary and defiant. Galois rose to propose a toast, his glass in one hand, his open pocket knife in the other. "To Louis-Philippe"—the King. His companions misunderstood the purpose of the toast and whistled him down. Then they saw the open knife. Interpreting this as a threat against the life of the King, they howled their approval. A friend of Galois, seeing the great Alexander [*sic*] Dumas and other notables passing by the open windows, implored Galois to sit down, but the uproar continued. Galois was the hero of the moment, and the artillerists adjourned to the street to celebrate their exuberance by dancing all night. The following day Galois was arrested at his mother's house and thrown into prison at Sainte-Pélagie.[49]

Dumas himself describes this event at length in his memoirs. Here is a portion:

> Suddenly, in the midst of a private conversation which I was carrying on with the person on my left, the name Louis-Philippe, followed by five or six whistles, caught my ear. I turned around. One of the most animated scenes was taking place fifteen or twenty seats from me.
> A young man who had raised his glass and held an open

dagger* in the same hand was trying to make himself heard. He was Evariste Galois, since killed by Pescheux d'Herbinville, a charming young man who made silk-paper cartridges which he would tie up with silk ribbons.†

Evariste Galois was scarcely 23 or 24 at the time. He was one of the most ardent republicans. The noise was such that the very reason for this noise had become incomprehensible.

All I could perceive was that there was a threat and that the name of Louis-Philippe had been mentioned: the intention was made clear by the open knife.

This went way beyond the limits of my republican opinions. I yielded to the pressure from my neighbor on the left who, as one of the King's comedians, didn't care to be compromised, and we jumped from the window sill into the garden.

I went home somewhat worried. It was clear this episode would have its consequences. Indeed, two or three days later, Evariste Galois was arrested.[50]

The amusing discrepancies between the two accounts are not entirely difficult to explain. Bell has taken his description almost word for word from Dupuy, who in turn has based his account on Dumas and the report in the *Gazette des Ecoles*.[51] The toasts Bell mentions as well as the description of the general atmosphere are found in Dupuy. But Bell has mistranslated: Dupuy writes that "... Dumas et quelques autres passaient par la fenêtre dans le jardin pour ne pas se compromettre ..."[52] which, in this context, means "... Dumas and several others jumped through the window into the garden in order not to be compromised." It does not here mean, "Dumas and several others passed by the open window in order not to be compromised." Did Bell ask himself why

---

* Literally, *poignard*.

† This is a literal translation of Dumas. We have not been able to discover exactly to what this occupation refers but it is a plausible guess that d'Herbinville made what the British call "crackers" (French, *diablotins*), party favors that pop when the ribbons are pulled and contain inspirational messages. They seem to have been invented at about this time.

Dumas should be passing by open windows in order not to be compromised? I do not know. He has also distorted the reason for the
banquet. Dupuy clearly states that it was a celebration for the acquittal of the nineteen conspirators.[53] But Bell has not mentioned
the trial. For consistency's sake (such as there remains), he must
therefore emphasize the obviously revolutionary character of the
gathering.

The issue of accuracy becomes more important when we question the most glaring omission in Bell's account: the absence of
any mention of Pescheux d'Herbinville. The single sentence in
Dumas is the only extant evidence that d'Herbinville was the man
who eventually shot Galois. Apparently Bell has not read Dumas
(otherwise he might have seen fit to close the discrepancies in the
banquet accounts—in order not be be compromised), but he does
claim to have read Dupuy who explicitly names d'Herbinville as
Galois' adversary.[54] Hoyle is guilty of the same charge; listing
Dupuy as a main reference, he relegates d'Herbinville to the ranks
of anonymous assassins. Infeld, who does identify d'Herbinville,
attempts to prove he was a police agent.

For mathematics, of course, it is not important to know exactly
who killed Galois; for historical accuracy it is. In light of the pleth-
ora of theories which have arisen to explain the cause of the celebrated duel, most of which involve police spies, agents provocateurs, and political overtones, the identity of d'Herbinville might
be a key piece of information. There are, in fact, two main candidates for the role of Galois' adversary, and all evidence indicates
that neither one was a police agent. Quite to the contrary. But for
now let us postpone conspiracy theories and return to Galois.

Galois was arrested at his mother's house the day following the
banquet, which does indicate that police or informers were at the
dinner, although the celebration was open to any subscriber. Galois was held in detention at Sainte-Pélagie prison until June 15,
when he was tried for threatening the king's life. Bell's description of this event is highly oversimplified. Indeed, the defense
lawyer did claim that Galois had actually said, "To Louis-Philippe, *if he betrays*," but that the noise had been such to drown
out the qualifying clause. Nevertheless, the matter took on a less

facetious appearance when the prosecutor asked Galois if he really intended to kill the king. Galois replied, "Yes, if he betrays." The prosecutor goes on to ask how Galois "can believe this abandonment of legality on the part of the king," and Galois answers, "Everything makes us believe he will soon turn traitor if he has not done so already." Galois is asked to clarify his remarks and basically repeats what he has already said: "I will say that the trend in government can make one suppose that Louis-Philippe will betray one day if he hasn't already."

As Dumas aptly remarks, "One understands that with such lucidity in the questions and answers, the discussion did not last long." Apparently moved by Galois' youth, the jury acquitted him within moments. Dumas goes on to say, "I repeat that this is a rude generation, perhaps a bit foolish, but you will recall Beranger's song *Les Fous*" ("The Fools" or "The Madmen").[55]

Shortly after the event, the Academy rejected Galois' memoir on the resolution of the equations, this time with Poisson as referee. The rejection was written on July 4, although according to Infeld Galois did not receive the letter until October, when he was in prison again.*[56] By this time, about eight months had passed since he had submitted the paper at Poisson's request. As we will see, Galois did not take the rejection lightly.

The cause for Galois' second arrest was preventative: on Bastille Day, July 14, 1831, he and his republican friend Duchatelet were apprehended dressed in Artillery Guard uniforms and heavily armed. Because the Artillery Guard had been disbanded on the last day of 1830 in fear of its becoming an instrument of the republicans, to wear the uniform was an outright gesture of defiance. It was also illegal. This was the charge brought against Galois, but not until the late date of October 23; he was sentenced to six months in prison. The sentence was confirmed by the court of appeals on December 3. In the meantime, Galois had been languishing in Sainte-Pélagie prison since his arrest in July.

* I have found no other source which either corroborates or contradicts Infeld's claim that the rejected manuscript was not received until October, three months after the actual rejection.

In Bell's fierce diatribe against this arrest he does not seem to comprehend that this was not the Paris of our day but Paris one year after a revolution, when street riots were rampant, assassination attempts not uncommon, and republican activity was dangerous.[57] The "celebration" Bell mentions was a republican demonstration on Bastille Day. Today such a demonstration would be considered patriotic; then it was seditious. This is exactly what the police chief decided when he went on record opposing the demonstration.[58] Bell concedes, "True, Galois was armed to the teeth when arrested, but he had not resisted arrest."[59] More precisely, Galois was carrying a loaded rifle, several pistols, and his dagger, a punishable offense even in our more moderate times. To say that he had not resisted arrest may also be inaccurate. The police came to Galois' house to detain him, but he had already decamped.

Galois' predicament was not helped by his friend Duchatelet, who drew on the wall of his cell a picture of the king's head lying next to a guillotine with the inscription, "Philippe will carry his head to your altar, O Liberty!"[60] Part of the delay in bringing Galois to trial was the fact that Duchatelet was tried first.

I am not arguing here for or against the justice of Galois' arrest. I am only trying to point out that he was behaving dangerously in a dangerous time. Two forces were clearly at work here: the government's intention to deal harshly with him after his threat of regicide and his own inability to keep out of trouble.

During his stay in prison, a number of events occurred which throw further light on Galois' personality. These incidents were recorded by the republican François Vincent Raspail. Raspail was an early botanist, one of the first to advocate the use of the microscope to examine cell structure in plants. He also had his troubles with the Academy and was sitting next to Dumas at the May 9 banquet. An ardent republican, he refused to receive the Cross of the Legion of Honor from Louis-Philippe and during the years 1830–1836 spent a total of twenty-seven months in prison.[61] Later in life, Raspail became a famous statesman. He lived to be about eighty and is now remembered by a boulevard and metro stop in Paris. One of his many arrests occurred at about the same time Galois was taken. Raspail recorded the following incidents in several of his letters. Infeld quotes him several times at great length but never explains who he was.

In a letter dated July 25, 1831, but conceivably revised for publication after Galois' death, Raspail wrote that his fellow prisoners had taunted Galois into drinking some liquor, a pastime at which he was apparently a novice:

> To refuse the challenge would be an act of cowardice. And our poor Bacchus had so much courage in his frail body that he would give his life for the hundredth part of the smallest good deed. He grasps the little glass like Socrates courageously taking the hemlock; he swallows it at one gulp, not without blinking and making a wry face. A second glass is not harder to empty than the first, and then the third. The beginner loses his equilibrium. Triumph! Homage to the Bacchus of the jail! You have intoxicated an ingenious soul, who holds wine in horror.[62]

The scene repeats itself. This time Galois empties a bottle of brandy in a single draft. Galois, drunk, pours out his soul to Raspail in what is either a retrospective invention or haunting prophesy:

> How I like you, at this moment more than ever. You do not get drunk, you are serious and a friend of the poor. But what is happening to my body? I have two men inside me, and unfortunately, I can guess which is going to overcome the other. I am too impatient to get the goal. The passions of my age are all imbued with impatience. Even virtue has that vice with us. See here! I do not like liquor. At a word I drink it, holding my nose, and get drunk. I do not like women and it seems to me that I could only love a Tarpeia or a Graccha.*

* The *Encyclopedia Britannica* states that Tarpeia, according to Roman legend, was the daughter of the Roman commander in charge of defending the capital against the Sabines. She offered to betray the citadel in exchange for what the Sabines wore on their left arms, that is, their bracelets. Taking her at her word, the Sabines crushed her beneath their shields. Graccha refers to Cornelia Graccha, the mother of Tiberius and Gaius, who is remembered as their educator as well as an accomplished author in her own right. Although hostile propaganda later suggested

And I tell you, I will die in a duel on the occasion of some *coquette de bas étage*. Why? Because she will invite me to avenge her honor which another has compromised.

Do you know what I lack, my friend? I confide it only to you: it is someone whom I can love and love only in spirit. I have lost my father and no one has ever replaced him, do you hear me . . .? [63]

The aftermath of the episode is neither heartwarming nor pleasant; Galois in a delerium attempts suicide:

We laid him out on one of our beds. But the fever of intoxication tormented our unhappy friend. . . . He would fall senseless only to raise himself with new exaltation, and he foretold sublime things which a certain reserve often rendered ridiculous.

"You despise me, you who are my friend! You are right, but I who committed such a crime must kill myself!"

And he would have done it if we had not flung ourselves on him, for he had a weapon in his hands. [64]

These are crucial paragraphs for the Galois legend and several points need to be made about them. Bell, in his account, says only, "Goaded beyond endurance, Galois seized a bottle of brandy, not knowing or caring what it was, and drank it down. A decent fellow prisoner took care of him until he recovered." [65] Thus the really important parts of the episode, which tell us something about Galois' character and which bear on future events, are omitted altogether.

Later, in attempting to understand the cause of Galois' death, Dupuy remarks, "If I credit an allusion of Raspail, Galois lost his virgin heart to *quelque coquette de bas étage*." [66] Bell writes, "Some worthless girl [*quelque coquette de bas étage*] initiated him." [67] Here Bell is taking a conjecture of Dupuy based on a letter of Raspail published seven years after Galois' death purporting to record an utterance of Galois spoken in a delerium a year before

she encouraged her sons' more revolutionary policies, she seems rather to have restrained them.

the duel as a characterization of real events. This can only be termed fabrication. And it is very likely that this piece of fabrication is responsible for the widespread belief that a prostitute was the cause of Galois' death.

Infeld, in his version of the prison scene, quotes the letters far more fully than Dupuy, but jumps from "Tarpeia and Graccha" to "Do you know what I lack, my friend?" In other words, he omits Galois' apparent prophecy that he will die in a duel. He also makes no comment whatsoever on Galois' suicide attempt. This selective presentation and slanting of evidence is characteristic of Infeld's book. He publishes any document or any portion of a document which does not interfere with his stated hypothesis that Galois was killed by the secret police. I will present more obvious examples later when I discuss the actual circumstances surrounding the duel.

On August 2, Raspail chronicles an interesting series of events which took place after his previous letters. On July 27, the prisoners were invited to attend a mass in memory of those killed during the July revolution a year earlier. Because many of the prisoners were political, the atmosphere was tense and an open riot was expected to erupt at any moment. A few prudent prison leaders defused the situation and two days passed without violence. At lock-up time on August 29, a shot was heard throughout the prison, followed by cries of "Help, murder!" The next day, the mystery was clarified. Raspail, quoting the conversation of another prisoner with the prison superintendent, writes:

> "Here are the facts, I am one of those in the attic room of the bathing pavilion. We were quietly going to bed. The man whose bed is between two casements had his face toward the window while undressing and he was humming a tune.
>
> "At that moment a shot was fired from the garret opposite. We thought our comrade was dead, but he was only unconscious. Not knowing where the shot had come from, nor how serious the wound was, we called for help. For in such a room, open in all directions through six windows, a better-aimed shot would have struck down its man."[68]

The shot, it turned out, came from a garret across the street where one of the prison guards lived. Galois was not the man who was at the window and wounded. However, he was in the same room and later thrown into the dungeon, evidently because he had insulted the superintendent, probably accusing him of having intentionally arranged the shooting. Raspail continues to record the conversation. The prisoner already quoted is talking:

"What? You have no order to seize the guilty man [the man who fired the shot]? But you have one to throw into the dungeon both the victim of this shameful trap and the witnesses of it? It may sound insolent to say that the administration pays turnkeys to murder prisoners. But what if this insolent statement is true? And I bear witness that no other insolence has come from those who were thrown into the dungeon. This young Galois doesn't raise his voice, as you well know; he remains as cold as his mathematics when he talks to you."

"Galois in the dungeon!" repeats the crowd. "Oh, the bastards! They have a grudge against our little scholar."

"Of course they have a grudge against him. They trick him like vipers. They entice him into every imaginable trap. And then too, they want an uprising."[69]

An uprising they got. This oblique conversation ends with the superintendent taking to his heels as the prisoners take control of the prison. The situation remains stalemated until late that night when the infantry is called in. The prisoners surrender without violence and remarkably no one is hurt.

I have tried to present this episode in as neutral a tone as possible. Infeld interprets the shot as an assassination attempt on Galois' life and later cites it in his Afterword as his first piece of evidence that Galois was murdered by the government.[70] I agree that the moderate government of Louis-Philippe would have liked to have been rid of all political extremists. But a conspiracy theory presumes that there exists a reason to single out a particular victim. Why Galois over Raspail? A shot was fired in a prison full of political prisoners on the verge of a riot, at night ("lock-up time"), into a room containing an unknown number of men, evidently "aimed" at someone else. Yes, it could have been an attempt to kill Galois. I do not find the evidence compelling.

More compelling is the evidence for the absolute hatred Galois had developed for the Academy, which I feel can only be termed paranoid. And, as is not uncommon with paranoids, there was a kernel of justification for the behavior. At some point in October, according to Infeld, Galois was notified of Poisson's rejection of his latest manuscript on the theory of equations.* Of this rejection Bertrand writes:

> Poisson decided to study the memoir; three months later he drew up a report that to [Galois] was a much too severe reproach.
>
> "We have made every effort," says Poisson, "to comprehend M. Galois' proof. His arguments are neither sufficiently clear nor developed for us to judge their rigor, and we are not in a position to even give an idea of them in this report."
>
> In declaring that, despite all efforts, he could not succeed in comprehending [Galois' work], Poisson's sincerity is very evident, and a reading of the memoir, printed twice since then, gives a sufficient explanation [to understand Poisson's failure]. The report ends with this benevolent remark:
>
> "The author claims that the proposition which is the subject of his memoir is part of a general theory rich in application. Often, different parts of a theory are mutually clarifying, and it is easier to understand them together than in isolation. One should rather wait for the author to publish his work in entirety before forming a definite opinion."
>
> Poisson refused to approve the proof, but he did not condemn it. In all fairness, he is irreproachable. He did as much as he could and was obliged to do.[71]

Bell, elaborating from Dupuy, states that Poisson found the manuscript "incomprehensible" but "did not state how long it had taken him to reach this remarkable conclusion."[72] I believe this is

* See footnote to page 168.

an unfair characterization of Poisson's comments. This is the rejection that, according to Bell, occurs before Galois' arrest.

In light of previous events and in light of his character, it is not terribly surprising that Galois reacted violently to what might nowadays be considered an encouraging rejection letter. He gave up all plans to publish his papers through the Academy and decided to publish them privately with the help of his friend Auguste Chevalier. Galois collected his manuscripts and in December, while still in Sainte-Pélagie, penned what must surely be one of the most remarkable documents in the history of mathematics, his *Préface*. The entire *Préface* runs about five pages. Infeld, to his credit, prints some of it, although he alters and omits certain parts at will. I here quote only the first page. The full text can be found in Bourgne and Azra:

> Firstly, you will notice the second page of this work is not encumbered by surnames, Christian names or titles. Absent are eulogies to some princes whose purse would have opened at the smoke of incense, threatening to close once the incense holder was empty. Neither will you see, in characters three times as high as those in the text, homage respectfully paid to some high-ranking official in science, or to some savant-protector, a thing thought to be indispensable (I should say inevitable) for someone wishing to write at twenty. I tell no one that I owe anything of value in my work to his advice or encouragement. I do not say so because it would be a lie. If I addressed anything to the important men of science or the world (and I grant the distinction between the two at times is imperceptible) I swear it would be thanks. I owe to important men the fact that the first of these papers is appearing so late. I owe to other important men that the whole thing was written in prison, a place, you will agree, hardly suited for meditation, and where I have been dumbfounded at my own listlessness in keeping my mouth shut at my stupid, spiteful critics: and I think that I can say "spiteful critics" in all modesty because my adversaries are so low in my esteem. The whys

and wherefores of my stay in prison have nothing to do with the subject at hand; but I must tell you how manuscripts go astray in the portfolios of the members of the Institute, although I cannot in truth conceive of such carelessness on the part of those who already have the death of Abel on their consciences. I do not want to compare myself with that illustrious mathematician but, suffice to say, I sent my memoir on the theory of equations to the Academy in February of 1830 (in a less complete form in 1829) and it has been impossible to find them or get them back. There are other anecdotes in this genre but I would be ungracious to recount them because, other than the loss of my manuscripts, those incidents do not concern me. Happy voyager, only my poor countenance saved me from the jaws of wolves. Perhaps I have already said too much for the reader to understand why, as much as I would have liked otherwise, it is absolutely impossible for me to embellish or disfigure this work with a dedication.[73]

The remainder of the *Préface* continues in much the same tone ("And thus it is knowingly that I expose myself to the laughter of fools"). Other of his writings are not dissimilar.[74] Among his papers is the picture of a bizarre, torsoless figure, captioned by Bourgne and Azra "Riquet à la Houppe."[75] The picture must have been drawn shortly before his death. It may be significant that Riquet à la Houppe was in French folklore a character, short, ugly, disdained by all, but nonetheless very clever.

6. THE DUEL AND THEORIES SURROUNDING IT

We are almost to the end of this short story. Galois remained in Sainte-Pélagie without further recorded incident until March 16, 1832, when he was transferred to the pension Sieur Faultrier. Ironically enough, this was to prevent the prisoners from being exposed to the cholera epidemic then sweeping Paris. Galois was due to be given his freedom on April 29. From this point on, the historical record is very scanty. On May 25, Galois writes to his friend Chevalier and clearly alludes to a broken love affair:

> My dear friend, there is a pleasure in being sad if one can hope for consolation; one is happy to suffer if one has friends. Your letter, full of apostolitic unction, has given me a little calm. But how can I remove the trace of such violent emotions as I have felt?
>
> How can I console myself when in one month I have exhausted the greatest source of happiness a man can have, when I have exhausted it without happiness, without hope, when I am certain it is drained for life?[76]

The letter continues in similar tones. Galois goes on to say that he is disgusted with the world: "I am disenchanted with everything, even the love of glory. How can a world I detest soil me?"[77]

The next few days are a complete blank. On the morning of May 30, the famous duel took place. The previous evening, Galois wrote several well-known letters to his republican friends:

> I beg patriots, my friends, not to reproach me for dying otherwise than for my country.
>
> I die the victim of an infamous coquette and her two dupes. It is in a miserable piece of slander that I end my life.
>
> Oh! Why die for something so little, so contemptible?
>
> I call on heaven to witness that only under compulsion and force have I yielded to a provocation which I have tried to avert by every means. I repent in having told the hateful truth to those who could not listen to it with dispassion. But to the end I told the truth. I go to the grave with a conscience free from patriots' blood.
>
> I would like to have given my life for the public good.
>
> Forgive those who kill me for they are of good faith.[78]

Galois also writes another, similar letter to two republican friends, Napoleon Lebon and V. Delauney:

My good friends,
I have been provoked by two patriots. . . . It is impossible for me to refuse.

I beg your forgiveness for not having told you.

But my adversaries have put me on my word of honor not to inform any patriot.

Your task is simple: prove that I am fighting against my will, having exhausted all possible means of reconciliation; say whether I am capable of lying in even the most trivial matters.

Please remember me since fate did not give me enough of my life to be remembered by my country.

I die your friend[79]

I will return to Galois' activities during the last night later. Now I want to dispose of some of the many theories which purport to explain the cause of this celebrated duel, beginning in each case with circumstances and eventually rising to facts.

If one is of a distrustful frame of mind, there is perhaps enough in the above two letters to raise suspicions of foul play. The attempts to make Galois the victim of royalists, a female agent provocateur, a prostitute, or a government conspiracy doubtlessly stem from these letters for there is no other direct evidence in existence. Thus we have the origin of Bell's assertion: "What happened on May 29th is not definitely known. Extracts from two letters suggest what is usually accepted as the truth: Galois had run afoul of political enemies immediately after his release."[80]

The first statement is accurate, the second is not. Dupuy certainly believes the exact opposite, as will become clear below. He does mention that Alfred Galois, unjustifiably in Dupuy's view, did maintain that his older brother was murdered. Because Bell "followed" Dupuy exclusively, one can only conclude that he took Alfred's position and termed it widely accepted or that he simply invented the whole thing.

Although Bell may have invented the theory, he is not its chief advocate. Infeld goes further. He assumes that the "infamous coquette'' was a female agent provocateur who set up Galois for the duel with a police agent. Infeld's evidence is by admission circumstantial. In addition to the bullet episode at Sainte-Pélagie, it consists of the following:[81] the police were known to have used spies; the police broke up a meeting of the Society of Friends of the People the night before Galois' funeral; Police Chief M. Gisquet wrote in 1840 that Galois "had been killed by a friend"; police spies were unmasked in 1848, at which time a claim appeared in a journal that Galois "had been murdered in a so-called duel of honor"; Galois' brother Alfred always maintained that Evariste had been murdered; Galois was abandoned by his adversaries and his seconds and found by a peasant.

Let me first counter circumstance by circumstance. Infeld's evidence is indeed consistent and does not contradict known facts. However, necessity does not follow from consistency. The bullet episode has already been discussed. It is true that the police used spies and that they were unmasked in 1848. I will return to this point shortly. Infeld does not mention that the newspapers announced Galois' funeral *before* the fact and explicitly named him as a member of both the Artillery of the National Guard and the Friends of the People. In any case, his membership in these organizations must have been widely known. One must weigh for oneself whether it is remarkable that the police knew of republican meetings. Infeld finds it suspicious that the police chief, eight years later, knew that Galois had been "killed by a friend." He does not find it suspicious that Dumas apparently knew more— precisely who that friend was. Dupuy feels that Alfred's position was the result of justifiable anger over his brother's death and points out some unlikely details Alfred attributed to the duel, such as stating that Evariste would have fired into the air. The assertion that Galois was abandoned to die, another of Alfred's claims, is also open to dispute. Dupuy notes that one of the witnesses went to Galois' mother the following day to explain what had happened.[82] He considers it more likely, then, that the witnesses were searching for a doctor when the peasant happened along. This explanation may be weak; nevertheless Infeld fails to point out that Mme. Galois was informed.

Let us now move from circumstance to harder evidence. As we have seen, Dumas' choice for Galois' opponent—which Infeld accepts—was a "delightful young man" named Pescheux d'Herbinville. More is known about him than his anonymity. He was, in fact, one of the nineteen republicans who were acquitted on charges of conspiring to overthrow the government. Is there any reason to suspect d'Herbinville was a police agent? The historian Louis Blanc, in his exhaustive *History of Ten Years*, writes:

> The trial gave rise to highly interesting scenes. In the sittings of the 7th of April, the president having reproached M. Pescheux d'Herbinville, one of the accused, for having [borne arms] and having distributed them, "Yes," replied the prisoner, "I have borne arms, a great many arms, and I will tell you how I came by them." Then, relating the part he had taken in the three days, he told how, followed by his comrades, he had disarmed posts, and sustained glorious conflicts: and how, though not wealthy, he had equipped national guards at his own cost. There still burned in the hearts of the people some of the fire kindled by the revolution of July; such recitals as this fanned the embers. The young man himself, as he concluded his brief defense, wore a face radiant with enthusiasm and his eyes filled with tears.[83]

In addition, Blanc mentions the appearance of General Lafayette during the trial: "The old general came to give his testimony in favor of the accused, almost all of whom he knew, and all saluted him from their places with looks and gestures of regard."[84]

D'Herbinville, it seems, was one of the heroes of the hour. After the acquittal, the crowd pulled his coach through the streets of Paris, "amid shouts of rapturous applause."

Bell (and Hoyle below), by not mentioning d'Herbinville at all, relieve themselves of the difficulty of explaining why Galois should be killed in a political duel with a fellow republican or why d'Herbinville should be considered a political enemy. Infeld is in a more difficult position. Having acknowledged d'Herbinville's existence, he must explain why neither Dumas nor Blanc, both republicans, nor evidently the extremely liberal Lafayette* (assuming he knew d'Herbinville personally), nor, one would gather from Blanc's account, any republican in Paris, ever held any suspicions that d'Herbinville was an agent. Infeld talks at length about the 1848 unmasking of the police spies but he does not mention the following extract from Dupuy:

> Pescheux was certainly not a "false-brother": all the men who acted as police agents during the reign of Louis-Philippe were revealed in 1848 when Caussidiere became chief of police, as witness Lucien de la Hodde.† If Pescheux were suspect, he would certainly not have been nominated as curator of the palace of Fontainebleau. It is absolutely necessary to discard the idea of police intervention and a framed assassination.[85]

We see that there are, even at this level, some serious difficulties with the political-enemies scenario. Infeld gets around these obstacles in characteristic fashion: in his bibliography he cites both Blanc and Dupuy as primary sources but *quotes neither*. In his Afterword he goes so far as to admit, "There is no reason to believe Pescheux d'Herbinville was a police agent." But then he goes on to say: "I believe there is enough circumstantial evidence to prove that the intervention of the secret police sealed Galois' fate. I do not believe it is possible to fit all the known facts without assuming Galois was murdered."[86]

While the reader is forming a rebuttal to this statement, we turn to events according to Hoyle, where we find an amusing inversion of Infeld's theory. Hoyle writes:

> Such are the bare bones of the story of the life and death of Evariste Galois. The classical biography of Galois [he then references Dupuy], in an attempt to add flesh to these bones, suggests that he was done to death by royalist enemies, as does E. T. Bell in his book *Men of Mathematics*. There are

---

* Lafayette had been considered republican enough to see his post of Commander of the National Guard dissolved after the events of December 1830.

† Hodde was a "republican" who was unmasked as a spy in 1848.

dark hints that the release from prison was but a device for encompassing his death, a necessary preliminary to his being matched against a highly skilled assailant in royalist pay. But why should Galois feel it critical to his honor that he should accept the challenge of a right-wing agent, especially if the agent were a known marksman? Gallic logic suggests on account of a girl.[87]

We first note the complete misrepresentation of Dupuy's position. Undeterred, Hoyle then goes on to dispose of the "infamous coquette" and propose his own theory:

> It is possible that the "infamous coquette" was the source of a purely personal quarrel, but it is the normal biological rule among mammals that sexual quarrels between two males cease as soon as one side seeks "accommodation." It is the normal rule that either party to such fights can simply walk away, which is just what Galois seems to have attempted to do.
>
> The more likely possibility is that Galois' habit of working mathematical problems in his head, his ability to think in parallel, caused serious animosities, and perhaps suspicions, to develop during his six months of imprisonment. There may have been suspicions that Galois was not wholly for the "cause," or even that he was an *agent provocateur*.[88]

I am divided between anger and hilarity. To suggest as Hoyle does that any republican in Paris suspected Galois after his expulsion from the Ecole Normale, his Artillery activities, his threat to the king, his arrests, trials, sentencings, resentencings, and prison activities borders on the fantastical. This is in addition to the fact that two or three thousand republicans later attended the funeral of this supposed agent provocateur. One might equally well claim Lenin had been suspected of being a Menshevik.

As to Hoyle's biosociological theories, he is contradicted by the (circumstantial) historical record. The greatest Russian poet, Aleksandr Pushkin, was killed in 1837 at the age of thirty-seven in a duel over his wife. England's Lord Camelford was killed in a duel over a prostitute. As late as 1838 members of the American legislature were engaging in similar duels. Toward the end of the eighteenth century, during election season, approximately twenty-three duels *per day* were fought in Ireland alone, unlikely just for political reasons. In the last decades of the nineteenth century, Paris newspapers carried notices of daily duels and their terms. These practices continued until World War I. The causes of such "affairs of honor" ranged from geese, to insults, to politics, to women.[89] Dupuy himself mentions that nothing was more common at the time in question than duels between republicans, and I think one can safely infer from his remarks that no one paid the slightest attention to them.[90]

But this rebuttal is as circumstantial as Hoyle's argument and not very satisfactory, so let us again rise to more concrete facts. The first of these is the existence of two fragmentary letters written to Galois by one Mademoiselle Stéphanie D., who is none other than the "infamous coquette" over whom the duel was fought. A prostitute? An agent provocateur? Most authors have assumed her identity to be an absolute mystery and that she, like Galois' opponent, is an anonymous casualty of history. Dupuy was apparently unaware of the letters or chose not to publish them. Bell and Hoyle never mention her name. Infeld calls her Eve Sorel (perhaps inspired by Stendhal). This is a strange state of affairs, for the letters are contained in Tannery's 1908 edition of Galois' papers. Tannery does not affix a name to the author of these fragments; it is left for the 1962 edition of Bourgne and Azra to attempt an identification. One can understand why Bell and Infeld did not mention her name since Tannery did not provide it. Hoyle does not have such an excuse, his book having been published in 1977. One cannot understand why these letters are never mentioned by anyone, especially by Bell and Infeld who cite Tannery as a major source for Galois' manuscripts.

The letters, as they exist, are copies made by Galois himself on the back of one of his papers.[91] The copies contain gaps, which may indicate that he had previously torn up the originals and could not completely reconstruct them. More likely, Galois purposely omitted any incriminating or personally distasteful segments. I say this because some words in the French versions are broken in half; one generally does not remember only half a word. Galois has certainly obliterated Stéphanie's last name in a fit of anger. Due to the fragmentary nature of these letters, their translation has proved difficult and may be uncertain in places. Where impossible to translate I have allowed the original French to stand.

Letter 1

Please let us break up this affair.   I do not have the wit to follow   a correspondence of this nature   but I will try to have enough to converse with you as I did before anything happened.   Here is Mr.   the *en   a   qui   doit   vous qu'a*   me and do not think about those things which did not exist and which never would have existed.

Mademoiselle Stéphanie D

14 May 183—

Letter 2

I have followed your advice and I have thought over what   has happened   on whichever denomination it may have happened between us. In any case, Sir, be assured there never would have been more. You're assuming wrongly and your regrets have no foundation. True friendship exists nearly only between people of the same sex, particular-ly   of   friends   full in the *vacuum* that   the absence of all feeling of this kind . . . my trust . . . but it has been very wounded . . . you have seen me sad   you have asked the reason: I answered you that I had sorrows that one had in-flicted upon me. I had thought that you would take this as anyone in front of whom one drops a word for these   one is not   The calm of my thoughts leaves me to judge the per-sons that I usually see without much reflection: this is the reason that I rarely regret having been wrong in my judgment of a person. I am not of your opinion *les sen   plus que   les a exiger   ni se*   thank you sincerely for all those who you would bring down in my favor.

These are highly tantalizing morsels, but is there anything else known about the author? Indeed there is. C. A. Infantozzi has examined the original of the first letter.[92] With the help of a magnifying glass and "appropriate lighting" he is able to discern Stéphanie's full signature under Galois' erasures: Stéphanie Dumotel. Further archival investigations by Infantozzi show she was Stephanie-Félicie Poterin du Motel, daughter of Jean-Louis Auguste Poterin du Motel, a resident physician at the Sieur Faultrier, where Galois stayed the last months of his life. In 1840 Stéphanie married Oscar-Theodore Barrieu, a language professor. Any presumption that she was a prostitute must at this point be discarded as a complete figment of Bell's imagination.

From Stéphanie's second letter it is not difficult to infer that Galois took some song of sorrows on her part too seriously and himself provoked the duel. On face value she certainly does seem to have been an unwitting participant in whatever transpired. Unfortunately, the establishment of Stéphanie's identity does not conclusively establish exactly what happened. But for those who still insist that agents provocateurs and right-wing assassins are not absolutely ruled out, I offer the following passage from André Dalmas' biography of Galois, in which he reprints an article from the June 4, 1832, issue of the Lyon "journal constitutionnel" *le Precurseur*:

> Paris, 1 June—A deplorable duel yesterday has deprived the exact sciences of a young man who gave the highest expectations, but whose celebrated precocity was lately overshadowed by his political activities. The young Evariste Galois, condemned for a year as a result of a toast proposed at a banquet at the Vendanges des Bourgogne, was fighting with one of his old friends, a young man like himself, like himself a member of the Society of Friends of the People, and who was known to have figured equally in a political trial. It is said that love was the cause of the combat. The pistol was the chosen weapon of the adversaries, but because of their old friendship they could not bear to look at one another and left the decision to blind fate. At point-blank range they were each armed with a pistol and fired. Only one pistol was charged. Galois was pierced through and through by a ball from his opponent; he was taken to the Hôpital Cochin where

he died in about two hours. His age was 22. L.D., his adversary, is a bit younger.[93]

It is frustrating that a report written only several days after the event is incorrect on the date of the duel, the date of Galois' death and on his age; this in itself should serve as a good lesson to armchair historians. And L.D.? The frustrations do not end. The man who fits the above description most accurately is not d'Herbinville at all but Galois' old friend Duchatelet who was arrested with him on Bastille Day. On the other hand, Bourgne and Azra give Duchatelet's first name as Ernest and d'Herbinville's surname also begins with a D, bearing in mind the variable orthographic procedures of the time (e.g., du Motel vs. Dumotel), not to mention the fact that he also participated in a political trial (though not with Galois). While I cast my vote for Duchatelet, the final distinction strikes me as unimportant. With Stéphanie's letters and the newspaper article we arrive at a very consistent and believable picture of two old friends falling in love with the same girl and deciding the outcome by a gruesome version of Russian roulette. This is my fairy tale. It has the virtues of simplicity and psychological truth. By comparison, the tales of Bell, Hoyle, and Infeld are baroque, if not byzantine, inventions.

Those who want to pursue the matter further are welcome to do so. I suggest a trip to Paris to check whether Duchatelet was boarding at the Sieur Faultrier during April 1832.

7. THE LAST NIGHT

We saw in the introduction to this essay how Bell all but states outright that Galois committed his theory of equations to paper the night before he was shot. James R. Newman repeats this as an assertion and the vision of the doomed boy, sitting by candlelight, feverishly bringing group theory into the world, seems to be the major myth which most scientists harbor concerning Galois. This is again due to Bell's embellishment of Dupuy who, in this instance, is sufficiently romantic of his own accord. But as has already been detailed at great length, Galois had been submitting papers on the subject since the age of seventeen. The term "group," used in the sense of "group of permutations," is found in all of them. During the night before the duel in addition to the letters already quoted, Galois wrote a long letter to his friend Chevalier.[94] He begins:

> My dear Friend,
> I have made some new discoveries in analysis.
> The first concerns the theory of equations, the others integral functions.
> In the theory of equations I have researched the conditions for the solvability of equations by radicals; this has given me the occasion to deepen this theory and describe all the transformations possible on an equation even though it is not solvable by radicals.
> All this will be found here in three memoirs.

Galois then goes on to describe and elucidate the contents of the memoir which was rejected by Poisson, as well as subsequent work. Galois had indeed helped to create a field which would keep mathematicians busy for hundreds of years but not "in those last desperate hours before the dawn." During the course of the night he annotated and made corrections to some of his papers. He comes across a note that Poisson had left in the margin of his rejected memoir: "The proof of this lemma is not sufficient. But it is true according to Lagrange's paper, No. 100, Berlin 1775."[95] Galois writes directly beneath it: "This proof is a textual transcription of that which we gave for this lemma in a memoir presented in 1830. We leave as an historic document the above note which M. Poisson felt obliged to insert. (Author's note.)" A few pages later, Galois scrawls next to a theorem: "There are a few things left to be completed in this proof. I have not the time. (Author's note.)"[96]

This famous inscription appears only once in Bourgne and Azra. It is unfortunate that Galois tarnished some of the romance by including his parenthetical "author's note." Galois ends his letter to Chevalier with the following request:

> In my life I have often dared to advance propositions about which I was not sure. But all I have written down here has been clear in my head for over a year, and it would not be in

my interest to leave myself open to the suspicion that I announce theorems of which I do not have complete proof.

Make a public request of Jacobi or Gauss to give their opinions not as to the truth but as to the importance of these theorems.

After that, I hope some men will find it profitable to sort out this mess.

I embrace you with effusion. E. Galois[97]

And that was the end. The funeral was to be held on June 2. During the previous evening, the police broke up a meeting of the Society of Friends of the People on the pretext that the republicans were planning a demonstration for Galois' funeral. Thirty of those present were arrested. The next day two or three thousand people were present at the services. Galois' body was interred in a common burial ground of which no trace remains today.

Later, Evariste's brother Alfred and his devoted friend Chevalier would laboriously recopy the mathematical papers and submit them to Gauss, Jacobi, and others. By 1843 the manuscripts had found their way to Liouville who, after spending several months in an attempt to understand them, became convinced of their importance. He published the papers in 1846.

There exist many fragments which indicate Galois carried on his mathematical researches not only while in prison, but right up to the time of his death. The fact that he could work through such a turbulent life is testimony to the extraordinary fertility of his imagination. There is no question that Galois was a great mathematician who developed one of the most original ideas in the history of mathematics. The invention of legends does not make him any greater.

8. HARSHER WORDS

The account of Galois' life given here is not entirely complete. There are more documents, letters, and events. No doubt I will shortly be exposed for having selectively presented evidence. The purpose of this essay, however, has not been one of completeness, nor entirely one of biography. No, the purpose has been to show that something is curiously out of sync. Two highly respected physicists and an equally well-known mathematician, members of the professions which most loudly proclaim their devotion to Truth, have invented history.

Bell's account, by far the most famous, is also the most fictitious. It is a myth devoid of such complications as a protagonist who is faulted as well as gifted. It is a myth based on the stereotype of the misunderstood genius whom the conservative hierarchy is out to conquer. As if the befuddled hierarchy is generally organized well enough for persecution. It is a myth based on a misunderstanding of the method by which a scientist works: as if a great theory could be written down coherently in a single night.

As an inventor of fairy tales, one can enjoy Bell; as a biographer it is unclear how far one can forgive him. Surely all his mistakes did not result from a poor knowledge of French. No, I believe Bell saw his opportunity to create a legend. The details which are absent in his account, such as Dumas at the banquet, such as d'Herbinville and the suicide attempt and Raspail, are those details which lend a concreteness and a humanness to Galois' life which a legend (at least a bad one) must not have. Unfortunately, if this was Bell's intent, he succeeded. After hearing of my investigation, physicists and mathematicians all open their conversations with me with the same question: "Did Galois really invent group theory the night before he was killed?" No, he didn't.

Infeld presents far more details. He is not interested in making Galois a legend. He does intend to make Galois a hero of the people. Politics is the guiding principle for Infeld. His book might be termed the proletarian interpretation of Galois; certainly parts of it read like the local Workers' Party publication. Infeld is very good at covering his tracks. To delete a phrase here, a paragraph there, a counterargument in between, is all that is necessary to create conspiracy from chaos.

As to Hoyle's motives, we can only take him at his word: he describes at length how as a child he was taught arithmetic by his mother, how he became proficient at mathematics, and how school for him became an excruciating bore. Hoyle was forced to learn to "think in parallel" in order to fool the teacher into believing he paid attention in class. He then tells us, "I mention these personal details because I believe they cast some light on the mysterious death of the French mathematician Galois." Further comment seems unnecessary.

Dupuy appears to have much less of a vested interest. I assume he included all the documents known to him at the time. If not, then he too should be scrutinized more carefully. He does seem a priori unwilling to accept a conspiracy theory.

At the very least, the three twentieth-century authors are guilty of distorting Dupuy's account and even falsifying it. In each case the story of Galois has been used to put a stamp of approval on the author's personal theories. Indeed, all history is interpretive. But if we do not approve, we understand the liberty: Galois, like Einstein, has passed into the public domain. No act or anecdote attributed to him is too outrageous to be given consideration. There is a closer analogy from farther afield. The Russian composer Reinhold Glière once wrote a symphony, his third, which ran well over an hour. Stokowski—the story goes—worked with Glière to edit the score down to manageable length. Since then every conductor presents his own edition. I do not know if I have ever heard the original.

The investigations of Galois discussed here have told us less about the man than about his biographers. The misfortune is that the biographers have been scientists. Because they appreciate his genius a century after its undisputed establishment, anyone who did not recognize it at the time is condemned. "In all the history of science," writes Bell, "there is no completer example of the triumph of crass stupidity over untamable genius." "Is it possible to avoid the obvious conclusion," asks Infeld, "that the regime of Louis-Philippe was responsible for the early death of one of the greatest scientists who ever lived?" The underlying assumption is apparent: Galois was persecuted because he was a genius, and all scientists, to a greater or lesser degree, understand that genius is not tolerated by mediocrity. A genius must be recognized as such even when standing drunk at a banquet table with a dagger in his hand. Anyone who does not recognize him becomes a fool, an assassin, or a prostitute. This is a presumption of the highest arrogance. Scientists should not be so enamored of themselves.

ACKNOWLEDGMENTS

I would like to thank Leonard Gillman for suggesting that I consolidate my research on Galois into this article. Thanks to Cecile DeWitt-Morette for the gracious gift of her time in translating, and most of all to Marc Henneaux for his translations, patience, discussions, and enthusiasm.

For this version I am grateful to René Taton for sending me some new information, which has changed a few of the facts, though not the conclusions, of the original *American Mathematical Monthly* version.

When the original version appeared, *Monthly* editor Ralph Boas would not allow me to thank him for his great help in supplying material and checking facts. I hope he will not mind if I acknowledge him now.

APPENDIX A

While most people responded favorably to my fairy tale, some did not. I reprint here an unsigned *New York Times* editorial (April 4, 1982) that appeared shortly after the *Scientific American* version of this article:

> On the last night of his 20-year-old life, before dying in a duel, the 19th century French mathematician Evariste Galois furiously scribbled outlines of the theories teeming in his mind, breaking off only to scrawl in the margin "I have not time." What he wrote in those last desperate hours, noted historian Eric Bell, "will keep generations of mathematicians busy for hundreds of years."
>
> Now this story has been a trifle deflated by an article in *Scientific American* which holds that Galois was not writing out new theories but merely redrafting an already written paper. The manuscript, an account of his celebrated theory of

groups (invaluable for solving Rubik's cube), had been returned to him for revision before publication.

Historical accuracy is a fine thing, but what a niggling cor-
rection to so haunting a story. The essence of the story, in
any event, is true: Galois was senselessly killed in his prime,
and he did spend his last night in desperate mathematical
work. Studious historians have already robbed us of Archimedes's "Eureka!" and Galileo's "But still (the earth)
moves." Let be Galois's "I have not time."

(Reprinted with permission.)

APPENDIX B
A Note on Galois Theory

In every high school algebra class we are asked to solve equations. Sometimes the equation might be very simple, such as
$x + 3 = 7$. Then we all know to subtract 3 from both sides to get
the result $x = 4$. A slightly more difficult sort of equation
is known as the quadratic, which is of the general form
$ax^2 + bx + c = 0$, where a, b, and c are real numbers. The quadratic, or second degree equation, is identified by the
x^2 term and every math student learns that its general solution is

$$x = \frac{-b \pm \sqrt{b^2 - 4ac}}{2a}.$$

For instance, if we have $7x^2 + 17x + 3 = 0$, the solution is

$$x = \frac{-17 \pm \sqrt{205}}{14}.$$

The important point is that the solution of a quadratic equation
generally requires the extraction of a square root, in this case
$\sqrt{205}$.

Similarly, a cubic equation, which is of the form $ax^3 + bx^2 +
cx + d = 0$ requires the extraction of a cube root. If you can take
an equation of the above form for any power of x and can solve it

by the operations of addition, subtraction, multiplication, division, and the extraction of roots, the equation is said to be solvable by radicals. (Another word for the $\sqrt{}$ sign is the "radical.")

All this may seem quite trivial but it is not. A solution by radicals to the general fifth degree equation, $ax^5 + bx^4 + cx^3 + dx^2 + ex + f = 0$, eluded investigators for centuries until the Italian mathematician Paulo Ruffini in 1799 and then the Norwegian Neils Henrik Abel in 1825 showed that such a solution does not actually exist.

Galois' contribution was to show under what conditions any equation of an *arbitrary* degree is solvable by radicals. This may seem like a limited achievement, but the techniques that Ruffini, Lagrange, Abel, and Galois invented to solve the problem, that is, the theory of groups, has applications far beyond the solution of equations. It has, in fact, become one of the most important branches of mathematics and virtually all modern theories of physics are based on it. Because this is a historical article, I am not going to give an exposition of Galois theory. The interested reader is referred to my *Scientific American* article (April 1982) for a few more details on that subject.

NOTES

NOTES TO ESSAY 3, "GEODESICS"

1. This history of domes and planetariums is largely based on information contained in Charles F. Hagar, *Window to the Universe* (West Germany: Carl Zeiss Company, 1980), available only from Carl Zeiss. Also, *Shelter* (Bolinas, Calif.: Shelter Publications, 1973), has proven useful.

2. Translation by G. R. Mair in *Callimachus, Lychophron and Aratus* by A. W. and G. R. Mair, Loeb Classical Library (Cambridge: Harvard University Press, 1955).

3. Franz Dischinger, "Fortschritte im Bau von Massivkuppeln," from *Der Bauingenieur*, vol. 10 (1925), pp. 362–366. Courtesy Dyckerhoff & Widmann, Munich.

5. *Shelter*, p. 111.

6. Steven Weinberg, *Gravitation and Cosmology* (New York: John Wiley & Sons, 1973), p. 147.

NOTES TO ESSAY 4, "ENTROPY"

1. E. Mendoza, ed., *Reflections on the Motive Power of Fire*, by Sadi Carnot and other papers (New York: Dover Publications, 1960), p. 19.

2. William Thomson (Lord Kelvin), "On the Dynamical Theory of Heat," in W. F. Magie, ed., *The Second Law of Thermodynamics* (New York: Harper Brothers, 1899), p. 118.

3. R. Clausius, *The Mechanical Theory of Heat* (London: John van Voorst, 1867), p. 117.

4. Clausius, *Mechanical Theory*, p. 133.

5. Ibid., p. 357.

6. Ibid., p. 365.

7. Jacob Landau, "The Necessity of Art in Education" (preprint), p. 13.

8. See, for example, Ilya Prigogine and Isabelle Stengers, *Order Out of Chaos* (New York: Bantam Books, 1984), p. 238.

9. Landau, "Necessity," p. 18.

10. Shannon's original 1948 papers are reprinted in Shannon and Weaver, *The Mathematical Theory of Communication* (Urbana: University of Illinois Press, 1949).

11. John R. Pierce, *An Introduction to Information Theory* (New York: Dover Publications, 1980), p. 24.

12. Paul Davies, *The Physics of Time Asymmetry* (Berkeley: University of California Press, 1977), p. 54.

13. Quoted in Emilio Segré, *From Falling Bodies to Radio Waves* (New York: W. H. Freeman, 1984), p. 242.

14. Leon Brillouin, *Science and Information Theory* (New York: Academic Press, 1962), pp. 164–167.

15. Leo Szilard, *Zeitschrift für Physik* 53 (1929): 840. Or see Pierce, *Information Theory*, pp. 199–207.

16. Or see Richard Feynman, *Lectures in Physics* (Reading, Mass.: Addison-Wesley, 1963), vol. 1, lecture 46, pp. 1–4.

17. Charles H. Bennett and Rolf Landauer, "The Fundamental Physical Limits of Computation," in *Scientific American* (July 1985), p. 48.

18. Daniel R. Brooks and E. O. Wiley, *Evolution as Entropy* (Chicago: University of Chicago Press, 1986).

19. Brooks and Wiley, *Evolution as Entropy*, p. 24.

20. Ibid., pp. 24-25.

21. T. Rothman, "The Seven Arrows of Time," *Discover* (February 1987), p. 76.

22. Brooks and Wiley, *Evolution as Entropy*, p. 27.

23. Ibid., p. 28.

24. Ibid., p. 41.

25. Ibid., p. 57.

26. Ibid., p. 63.

27. Ibid., chap. 4.

28. Ibid., p. 63.

29. Richard P. Feynman, *"Surely You're Joking, Mr. Feynman"* (New York: W. W. Norton, 1985), last chapter.

30. Brooks and Wiley, *Evolution as Entropy*, p. 300.

31. Jeremy Rifkin, *Entropy: A New World View* (New York: Viking Press, 1980).

32. Jeremy Rifkin, *Algeny: A New Word—A New World* (New York: Viking Press, 1983).

33. Rifkin, *Entropy*, p. 33.

34. Ibid., pp. 40–41.

35. Ibid., p. 38.

36. Ibid., p. 246.

37. Ibid., p. 235.

38. Ibid., p. 260.

39. Ibid., p. 206.

40. Ibid., p. 254.

41. Lionel Harrison, "Biology Syncretized," *Science* 232 (1986): 1027.

42. Jolande Jacobi, ed., *Paracelsus, Selected Writings* (Princeton: Princeton University Press, 1973), p. 39.

NOTES TO ESSAY 6, "GALOIS"

1. Freeman Dyson, *Disturbing the Universe* (New York: Harper and Row, 1979), p. 14.

2. E. T. Bell, *Men of Mathematics* (New York: Simon and Schuster, 1937), p. 375.

3. James R. Newman, *The World of Mathematics* (New York: Simon and Schuster, 1956), vol. 3, p. 1534.

4. *Checklist of the Bullitt Collection of Mathematics* (Louisville, Ky.: University of Louisville, 1979).

5. John Sommerfield, *The Adversaries* (London: Heinemann, 1952). I thank H. Schwerdtfeger for calling my attention to this book.

6. Leopold Infeld, *Whom the Gods Love: The Story of Evariste Galois* (New York: Whittlesey House, 1948).

7. Fred Hoyle, *Ten Faces of the Universe* (San Francisco: W. H. Freeman, 1977), chap. 1.

8. Paul Dupuy, "La Vie d'Evariste Galois," *Annales de l'Ecole Normale* 13 (1896): 197–266.

9. Bell, *Men of Mathematics*, p. vii.

10. Jules Tannery, ed., *Manuscrits d'Evariste Galois* (Paris: Gauthier-Villars, 1908).

11. Robert Bourgne and J. P. Azra, eds., *Ecrits et mémoires mathématique d'Evariste Galois: Edition critique integrale de ses manuscrits et publications* (Paris: Gauthier-Villars, 1962).

12. Alexandre Dumas, *Mes mémoirs* (Paris: Editions Gallimard, 1967), vol. 4, chap. 204.

13. François Vincent Raspail, *Lettres sur les prisons de Paris* (Paris: 1839), vol. 2.

14. Bell, *Men of Mathematics*, p. 362.

15. Ibid., p. 363.

16. Dupuy, "La Vie," p. 203.

17. Bell, *Men of Mathematics*, p. 364. Compare with Dupuy, "La Vie," p. 205.

18. Dupuy, "La Vie," pp. 255–256.

19. Ibid., pp. 254–255.

20. Ibid., p. 256.

21. Ibid., p. 209; Bell, p. 368.

22. Infeld, *Whom the Gods Love*, p. 306.

23. René Taton, "Sur les relations scientifiques d'Augustin Cauchy et d'Evariste Galois," *Review d'Histoire des Sciences* 24 (1971): 123.

24. Bell, *Men of Mathematics*, p. 368.

25. Ibid., p. 369.

26. Joseph Bertrand, "La Vie d'Evariste Galois par Paul Dupuy," *Eloges Académique* (Paris), new series (1902), pp. 329–345.

27. Bell, *Men of Mathematics*, p. 371.

28. Dupuy, "La Vie," p. 217.

29. Bourgne and Azra, *Ecrits*, p. 27. See also pp. 21–25.

30. I assume here, as elsewhere, the chronology given by Bourgne and Azra, *Ecrits*, pp. XXVII–XXXI.

31. Bell, *Men of Mathematics*, p. 370.

32. Taton, "Les relations," p. 134.

33. Ibid., p. 139.

34. Bourgne and Azra, *Ecrits*, p. XXVIII.

35. Taton, "Les relations," p. 138.

36. See, for example, Lilian Lieber, *Galois and the Theory of Groups* (1932).

37. Bourgne and Azra, *Ecrits*, p. XXVIII.

38. Bell, *Men of Mathematics*, p. 366.

39. Dupuy, "La Vie," p. 221, and Bourgne and Azra, *Ecrits*, p. XXIX.

40. Bourgne and Azra, *Ecrits*, p. 462. Also Dupuy, "La Vie," p. 225. Translated in part by Infeld, *Whom the Gods Love*, p. 155.

41. Dupuy, "La Vie," pp. 227–228.

42. This account is based on information from Louis Blanc, *History of Ten Years* (London: Chapman and Hall, 1844). A consistent account, also based on Blanc, is given by Infeld, *Whom the Gods Love*, chap. 5.

43. Dupuy, "La Vie," p. 234.

44. Bell, *Men of Mathematics*, p. 372.
45. C. Henry, "Manuscrits de Sophie Germain," *Revue Philosophique* 8 (1879): 631.
46. See, for example, his Afterword.
47. Infeld, *Whom the Gods Love*, p. 169.
48. Dumas, *Mes mémoirs*, p. 331.
49. Bell, *Men of Mathematics*, p. 372.
50. Dumas, *Mes mémoirs*, pp. 332–333.
51. Dupuy, "La Vie," pp. 234–235.
52. Ibid., p. 235.
53. Ibid., p. 234.
54. Ibid., p. 247.
55. Because of a change of libraries, this account of the trial is based on a different edition of Dumas' memoirs: Alexandre Dumas, *Mes mémoires* (Paris: Union Générale d'Editions), vol. 2, chap. 37 (no copyright date given).
56. Infeld, *Whom the Gods Love*, p. 230.
57. See, for example, T.E.B. Howarth, *Citizen King: The Life of Louis-Philippe* (London: Eyre & Spottiswoode, 1961).
58. Dupuy, "La Vie," p. 238.
59. Bell, *Men of Mathematics*, p. 378.
60. Dupuy, "La Vie," p. 238.
61. Dora B. Weiner, *Raspail: Scientist and Reformer* (New York: Columbia University Press, 1968).
62. Raspail, *Lettres*, p. 84.
63. Ibid., p. 89.
64. Ibid., p. 90.
65. Bell, *Men of Mathematics*, p. 374.
66. Dupuy, "La Vie," p. 245.
67. Bell, *Men of Mathematics*, p. 374.
68. Raspail, *Lettres*, pp. 117–118.
69. Ibid., p. 118. Also discussed by Dupuy, "La Vie," p. 243.
70. Infeld, *Whom the Gods Love*, p. 308.
71. Bertrand, "La Vie," pp. 340–341. (He does not give a reference for the source of Poisson's letter.)
72. Bell, *Men of Mathematics*, p. 371.
73. Bourgne and Azra, *Ecrits*, pp. 3–11.
74. See again ibid., pp. 21–27.
75. Ibid., facsimiles.
76. Ibid., pp. 468–469.

77. Ibid., p. 469.

78. Ibid., p. 470.

79. Ibid., p. 471.

80. Bell, *Men of Mathematics*, p. 375.

81. Infeld, *Whom the Gods Love*, pp. 308–311.

82. Dupuy, "La Vie," pp. 247–248.

83. Blanc, p. 431.

84. Blanc, *History*, p. 431.

85. Dupuy, "La Vie," p. 247.

86. Infeld, *Whom the Gods Love*, p. 310.

87. Hoyle, *Ten Faces*, p. 14.

88. Ibid., p. 15.

89. See, for example, Charles Mackay, *Extraordinary Popular Delusions and the Madness of Crowds* (New York: Noonday Press, 1932), chap. on "Duels and Ordeals"; and Roger Shattuck, *The Banquet Years* (New York: Vintage Books, 1968), chap. 1.

90. Dupuy, "La Vie," p. 247.

91. The letters and descriptions of them are in Bourgne and Azra, *Ecrits*, pp. 489-491.

92. C. A. Infantozzi, "Sur la mort d'Evariste Galois," *Revue d'Histoire des Sciences* 21 (1968): 157.

93. André Dalmas, *Evariste Galois, Révolutionnaire et Géometre* (Paris: Fasquelle, 1956), pp. 77–78.

94. Bourgne and Azra, *Ecrits*, p. 173.

95. Ibid., p. 48 and facsimiles.

96. Ibid., p. 54 and facsimiles.

97. Ibid., p. 185.

INDEX

Abel, N. H., 176, 193
Ackerman, T., 133
action and reaction, 49
Aftermath, The, special issue of
  Ambio reprinted, 110
Albrecht, A., and new inflation, 26,
  36, 41
Aleksandrov, V., 112, 121–22, 131,
  142, 145–46
Aleksandrov, V., and G. Stenchikov,
  confirmatory nuclear winter study,
  112
Alfvén, Hannes, 3
Ambio, special issue on nuclear war,
  110, 116
Amicus, Ehrlich in, 114, 131, 145
Ampère, A. M., 157n
Anderson, P., 119
anisotropic models, 13–15; quantum
  damping of anisotropy in, 15, 38
anthropic principle, 17n, 32, 49
Aratus, 52–53
Archimedes, 66
arrow of time, and second law of
  thermodynamics, 82
Atwood, W., 53
Ayres, R. U., and Hudson Institute on
  nuclear war, 114–15, 145

bandwagons, rolling on, vii–viii
Barrow, J., 17n
Batten, E. S., and Rand study of
  nuclear war, 114–15, 125, 145
Bauersfeld, W., inventor of geodesic
  dome, 56–59, 63, 74
Bell, E. T. (*Men of Mathematics*),

biography of Galois critiqued, 148–
  90, 197–99
Bennett, C., and logically reversible
  computers, 96
Bertrand, J., on Galois, 156, 174,
  198–99
Big Bang: in standard model, ix, 4–7
Big Stop, 24
black-body radiation, 94
Blanc, L., 180–81, 198, 200
Boltzmann, L., and statistical
  mechanics, 83–90
Bondi, H., 26
bounce, 24
Bourgne, R., and J. P. Azra, complete
  edition of Galois' papers, 151,
  175–76, 183, 186–87, 197–200
Bourgogne, Vendanges des
  (restaurant), 164–66
Brans-Dicke theory, 24
Brillouin, L., and Maxwell's Demon,
  94, 103, 196
Brooks, D. (coauthor of *Evolution as
  Entropy*), 97–101, 196
Brundit, G. B., 12
Bullitt archives of mathematics, 149,
  197

Canuto, V., 23
Carnot, L., 78
Carnot, S., 78–80, 195
Carr, B., 17n, 19
Carrier, G., 131–33
Cauchy, A., 155, 157n, 158–60
Center for Energy and Environmental
  Studies (CEES), at Princeton, 118–
  19